EVERYTHING IS SOMEWHERE
IN THE U.S.A.

Also by Jack McClintock and David Helgren, Ph.D.

Everything Is Somewhere: The Geography Quiz Book

EVERYTHING IS SOMEWHERE IN THE U.S.A.

Jack McClintock and David Helgren, Ph.D.

QUILL

William Morrow

New York

Library of Congress Cataloging-in-Publication Data

McClintock, Jack.
    Everything is somewhere in the U.S.A. : the geography quiz book / Jack McClintock and David Helgren.
      p.   cm.
    ISBN 0-688-12733-9
    1. Geography—Miscellanea.   I. Helgren, David M., 1947–
II. Title.
G131.M36   1994
917.3—dc20   93-49458
                                                 CIP

Printed in the United States of America

3 4 5 6 7 8 9 10

BOOK DESIGN BY RICHARD ORIOLO

CONTENTS

CONTENTS

Preface

When *Everything Is Somewhere* came out in 1986, we had no idea it would stick around so long, let alone demand a kind of sequel. We wrote *Everything Is Somewhere* with the idea of drawing attention to the lamentable want of geographical knowledge in America, and to prove that learning about our world can be a pleasure. We're glad it has helped to do that.

The project began when David Helgren inflicted a modest geography quiz on his students at the University of Miami, where he then taught. A third of them could not locate France or the South Pacific on a world map—and 7 percent couldn't find Miami, which was just outside the window. These were students from all over America, and Helgren's little quiz made national news and became a symbol of our geographic illiteracy. The ensuing publicity helped lead to the National Geographic Society's establishing a foundation for geography in the schools, now expanded to include a Geographic Alliance in every state (with Helgren a founder of the one in California, where he now teaches) and a dramatic expansion of geography education.

Today, with the world splintered into ever more complex and mysterious fragments, there is even greater reason to understand the shape and meaning of the whole.

And the best place to start is at home.

EVERYTHING IS SOMEWHERE
IN THE U.S.A.

Introduction

Don't you get a thrill at the shopping mall's information kiosk when you see that little red arrow on the glass-topped pedestal map: YOU ARE HERE? What could be more satisfying in life than knowing where you stand?

There's a lot more to geography than state capitals, major mountain ranges, headhunters, topless "natives," and foldout maps in old issues of that slickly colorful, evocative, yellow-bordered magazine. Geography is a way of organizing all knowledge, analyzing spatial relationships, and figuring out how one thing changes another, thus arriving at some understanding of what the world is like and how it got that way. By looking at things geographically, we structure our discoveries and reintegrate a mysterious world that we usually apprehend only in parts.

And what a world. The more you see, the more your sense of wonder grows. How to keep track of it all?

That's where those beautiful foldout maps come in, not to mention atlases, globes, and navigational charts. We think there are two kinds of people: those who love maps, and those who fear and loathe them. Map-fearers get lost in maps because they're afraid of them—after all, a map can lead you off the beaten path as well as along it. Map-lovers know that and think it's terrific. Maps are for exploration as well as for finding your way home.

The geography of the individual begins at home. As you know, this is a physical place as well as an emotional, social, and sentimental one. Home is where one returns after exploring, to ponder what one has seen and rest up safely for the next expedition.

Outside the home, geographers see concentric geographical "shells" through which we travel or from which we acquire stuff—which geographers call resources. A resource is anything you think you need, whether it's a cup of coffee, a dollar, a lesson, or a dream. These shells look like a target with a bulls-eye—home—in the center. In the next shell is your community of immediate interaction: friends, schoolmates, colleagues, tradespersons, the operators of shops and services. Then comes your region or state—delineated by political boundaries. The next shell is your country, and the next is the world.

Each shell has its own set of locations, places, and relationships. We can zoom in close to examine geographical questions in the immediate neighborhood. How many bookshops can a certain kind of city support? Can a city with other characteristics support more? What happens when you add an espresso bar and live guitar music on Friday nights? Does it matter what kind of guitar music? Or we can pull back and check out the whole country, as we do a little in this book. Ask anything. What kind of transportation networks has the United States used? How have they evolved? What physical and cultural factors work against the success of mass transit? How has air travel changed since 1980s' deregulation? Is it more like bus travel used to be? What do sports cars, customized cars, pickup trucks, step vans, or Jeeps tell you about the kinds of neighborhoods you see them parked in? What does your car say about your neighborhood? (What do your neighbors say about your car?) Why do so many city dwellers drive Jeeps?

And the maps? The oldest are mental maps, the ones you'd better not leave home without. They are like little computer graphics in your head that shift and revolve and advance as you move from place to place, taking in and recording the distance traveled, the look of the neighborhood, the slant and intensity of the light, smells in the air—all to help guide you, tell you where you are, and let you know when you've arrived. Where would you be without these mental maps? You'd be a certified mental incompetent without them, truly. You'd be *lost*.

To be formal about it for a moment, geography has five major themes or elements. They are location, place, interaction, movement, and region.

Location is where a thing is, in either absolute or relative terms—its latitude and longitude, or its direction and distance from someplace else. As we will see throughout the United States, few locations are accidental. Locations are chosen, or they happen for reasons.

Place is what is at the location—your house, Alcatraz, Harvard Yard. A place is created by nature or culture or both, and these vary over time and space, so that each place has qualities uniquely its own. No one would ever mistake one place for another even if they somehow switched locations. Try putting Alcatraz where Harvard Yard is in your mind, and see.

Interaction is the way one thing affects another across space and time. Environments can shape people; people change environments. Two examples of interaction: sharing and stealing. (Think about how those around you might change in reaction to these interactions.)

Movement is about transportation, migration, and communication—the incessant creep, flow, discharge, tote, fetch, and carry of things and ideas from one place to another. Movement is about distance, and thus hidden within the idea of movement is the cost of distance, which naturally varies. It costs more to move things away from the earth's surface than toward it, whether we're leaping away in a rocket or burrowing inside with a drill bit.

As for *regions*, these are to geographers what eras are to historians: a way of dividing things up so as to better understand them, bit by bit (not that, as geographers, we won't eventually reassemble them in order to grasp the whole). For example, the Corn Belt and the Pacific Northwest are regions. We can grasp them as somehow consistent within themselves yet different from each other. And both are part of North America, which itself would be different without either one.

Pondering all this, we can see our tiresome old situations in a fresh, sometimes even startling, way. One of us, for example, thinking about where he lives—Miami—suddenly saw clearly that the main reason his hometown is so hideously troubled by crime, racism, and epidemic incivility is less because it is a "bad" place, as so many, including himself, have often thought, but because it is a "good" place. It is desirable. It is happily located for business and travel in any direction, abroad or at home; the weather is warm and bright; the air is clean—you can sleep outdoors if you have to. There are always coconuts to eat, even if they're much too high in saturated fat. By contrast to a dingy-brick, dying Rust Belt town in the north, or one of those corrupt, polluted, violent spots that unfortunately flourish to the south, it's paradise. That's why it's full of northerners looking for jobs, and Haitians and Cubans and Hondurans and Nicaraguans and Guatemalans hoping only to be safe and to prosper. For good or ill, or both, geography has made Miami what it is.

History would be unintelligible if things didn't happen at particular locations, places, regions. Knowing geography is our best chance to put it all back together, figure it out, and come to know where we stand.

The Whole U.S.A.

America will be "completed" in about a hundred years.

Our population, now approaching 260 million, will continue to grow until it reaches a little over 300 million in the year 2040. Then, according to the U.S. Census Bureau, it will begin a slow decline until it levels off at around 290 million in the year 2080. We— or rather they—will live mostly in cities, and mostly within a hundred miles of an ocean or major lake.

These are the patterns of settlement already established, and scholars say it takes very big disruptions to change them—bigger, as we have observed in California, Florida, and the Midwest, than any little earthquake, hurricane, or flood. But most of the great population growth will take place in the *next* few decades. The decisions of those citizens will determine the future shape of the nation. As the selection of questions here suggests, we are a diverse people in a variable land. And, of course, our geography will never be completed; we Americans love change too much.

1. A famous American novelist has written what might be called geographical fiction about such places as Hawaii, Texas, the Chesapeake Bay, and the South Pacific. Who is this literary geographer?

2. How far did you travel during the last hour, while you were just sitting there reading this book?

3. You know that Russia is the world's largest country by far. Can you rank these four countries by area?

   Brazil _____

   Canada _____

   China _____

   United States _____

4. What proportion of Americans believed, in late 1993, that Elvis Presley was still alive?

5. Which is farthest north?

   Portland, Maine _____

   Portland, Oregon _____

   Portland, Michigan _____

   Portland, Texas _____

6. What is a GIS?

7. How much time does the average American spend waiting in lines during a lifetime?

8. What is the world's third-largest country in population, behind China and India?

9. Where is the geographical center of *North America*? (Hint: It has a sporty name.)

10. Where is the only royal palace in the United States?

11. How many time zones are there in all fifty states?

12. What are these: Biscayne, Bel Air, Monterey, Montclair, Bonneville, El Dorado, Laredo, New Yorker, Scottsdale, Cheyenne, Laguna, Malibu?

13. How many Americans live alone?

14. In the early 1960s, an American over sixty-five had an average of 7.2 teeth left. How many did an American have thirty years later?

15. If you put all fifty states inside a rectangle of latitude and longitude, where would the edges be?

16. In 1959, a single father headed one of every 100 families. How many were headed by single fathers in the early 1990s?

> One in ten _____

> One in twenty-five _____

> One in eighty-five _____

> One in ninety-five _____

17. In which set of American cities do the residents walk the fastest?

> New York, Buffalo, and Boston _____

> Los Angeles, Sacramento, and Shreveport _____

18. How many American children witness domestic violence each year?

> About 1 million _____

> About 3 million _____

> About 11 million _____

19. About what percentage of American fathers who are divorced from or never married the mother of their children support their children?

20. What percentage of American women will become widows?

21. Why were American houses dirtier in the mid-1980s than twenty years before?

22. One fourth of American commuters eat breakfast in the same place. Where is it?

23. What is America's favorite fruit? How much of it do we eat?

24. What federal agency controls more U.S. land than any other?

25. If current trends continue, in what year will the Hispanic population of America surpass the black population?

> 1997 _____

> 2000 _____

> 2013 _____

> 2027 _____

> 2057 _____

26. Which of the contiguous states has the most miles of shoreline? And what's the difference between "shoreline" and "coastline"?

27. How many people are hungry—a state defined as "the chronic underconsumption of nutrients"—in America?

> About 1 million _____

> About 2 million _____

> About 5 million _____

> About 30 million _____

28. The average age in America, in 1990, was 36.5. Which county in America has the oldest average population?

> Llano, Texas _____

> Gwinnett, Georgia _____

> Monterey, California _____

> Polk, Florida _____

29. Which county has the youngest average population in America?

> Dade County, Florida _____

> Chattahoochee County, Georgia _____

> Wade Hampton County, Alaska _____

30. In the later 1950s, there were over 4,000 drive-in movie theaters in the United States. How many are left?

> About 2,000 _____

> About 3,200 _____

> About 200 _____

> About 700 _____

> About 1,000 _____

31. Where does America rank in the world in average income?

Number one _____

Number two _____

Number three _____

Number ten _____

Number eleven _____

Number twelve _____

Number thirteen _____

32. Where does America rank in the world in life expectancy?

Number one _____

Number two _____

Number three _____

Number ten _____

Number eleven _____

Number twelve _____

Number thirteen _____

33. What proportion of Hispanic-Americans carry a major credit card?

About 16 percent _____

About 36 percent _____

About 66 percent _____

About 86 percent _____

34. What proportion of African-Americans carry a major credit card?

About 5 percent _____

About 15 percent _____

About 25 percent _____

About 35 percent _____

About 45 percent _____

35. In the modern art section of the Metropolitan Museum in New York, what proportion of the *artists* represented are female?

36. In the modern art section of the Metropolitan Museum in New York, what proportion of the *nudes* represented are female?

37. Where in the United States might you find 300 globes, 4 million maps, and 53,000 atlases from around the world—all in one place?

38. What is the difference between a "shaded relief" and a "raised relief" map?

39. The United States has what proportion of the world's lawyers?

40. What do you call a person who speaks three languages? What do you call a person who speaks two languages? What do you call a person who speaks one language?

41. What percentage of American homes have computers?

About 11 percent _____

About 22 percent _____

About 33 percent _____

About 44 percent _____

About 55 percent _____

42. What percentage of the American people believe it is possible that the Holocaust, Nazi Germany's extermination of 6 million Jews and millions of others, never happened?

43. How many steps does an American adult take in a day, on the average?

44. Did Christopher Columbus discover America?

45. Where is the first permanent (so far) colony on the land of the United States?

46. In a plate tectonic sort of way, in which direction is the United States moving?

47. How many acres in a square mile?

64 _____

620 _____

6,400 _____

48. What is the westernmost point in the lower forty-eight states?

49. What is the southernmost city in the lower forty-eight states?

50. What is the easternmost point in the forty-eight coterminous states?

51. Where is the northernmost point in the forty-eight coterminous states?

52. The commercialization of what famous American vacation spot led to the idea that the national parks should be protected from tacky development?

53. In 1789, a man named Christopher Colles did something significant in U.S. travel history. What was it?

54. The "World's Longest Graveyard" is in the United States. Can you give another of its names?

55. Where was the world's first motel?

56. What do these entertaining white males all have in common: Bud Abbott, Charles Addams, Robert Blake, David Copperfield, Lou Costello, Brian De Palma, Danny DeVito, Art Garfunkel, Jack Nicholson, Paul Simon, Bruce Springsteen?

57. Match these national parks with their outstanding environmental characteristics:

Acadia	volcanic eruptions, big bears
Big Bend	coral reefs
Everglades	rugged seashore
Mesa Verde	mid-latitude rain forest
Sequoia	geysers and hot springs
Zion	great trees and mountains
Yellowstone	ancient Indian cliff dwellings

Olympic	desert mountains and river
Katmai	subtropical wetland
Biscayne	multicolored gorges

58. John Steinbeck wrote a novel entitled *East of Eden*, which took place largely in the town of his birth, which was———.

59. The largest state in area is Alaska, the smallest is Rhode Island. But which state is precisely in the middle: twenty-fifth largest in area?

Georgia _____

Illinois _____

Iowa _____

Delaware _____

60. What was the largest addition of land ever made at once to the United States?

61. Where did the characters in Walt Kelly's comic strip *Pogo* live?

62. If you were to measure the width of the United States, where would you put the ruler and how wide would it be?

63. What is America's northernmost state?

64. What is America's westernmost state?

65. What is America's easternmost state?

66. Where is the highest waterfall on the North American continent?

67. Canada, the United States, and Mexico are all in North America. Two of these countries, but not the third, are members of the same "culture realm." Name the two.

68. What is the world's largest single source and clearinghouse of information?

69. Why does an Appalachicola oyster reach market in two years, a Chesapeake in three, a Long Island in four or five, and a Cape Cod in six?

70. What is the average population density of the United States?

      6 persons per square mile _____

      11.5 persons per square mile _____

      70.3 persons per square mile _____

      111 persons per square mile _____

71. Are we still, as Vance Packard suggested in his book of the same title, *A Nation of Strangers*?

Answers

The Whole U.S.A.

1. James Michener.

2. You were carried about 66,000 miles, but you scarcely noticed because your chair, cup of coffee, the cat, everything around you, was moving along with you. This is how far the earth travels per hour on its orbit around the sun.

3. Russia has 6.5 million square miles. Canada has 3.83 million square miles; China has 3.69 million square miles; and the United States has 3.67 square miles, making it the fourth-largest country in the world. Brazil has 3.28 million square miles.

4. Seven percent of Americans thought Elvis was still alive.

5. At slightly more than 45 degrees north latitude, it is Portland, Oregon.

6. Geographic Information Systems, usually shortened to GIS, is the new electronic geography in which computers store maps, aerial photos, and satellite images in multiple layers, permitting immediate mapping of multiple geographic variables in multiple places. GIS courses, as well as increasing concern about geographic education, are giving college geography programs new life, and GIS systems are on the budget of many government agencies and large landowning companies these days.

7. The average American waits in line for five years.

8. The United States is the third most populous country in the world, with a population growing past 250 million.

9. Rugby, North Dakota, which is 1,500 miles from the Atlantic Ocean, the Pacific Ocean, and the Gulf of Mexico.

10. The Iolani Palace was the official residence of the Hawaiian monarchy.

11. There are six time zones: Eastern, Central, Mountain, Pacific, Alaska, Hawaii-Aleutian.

12. American places with cars named after them.

13. About 22 million live alone. One in four households are single households, according to the Census Bureau. Women are more likely to live alone than men, outnumbering them three to two. Women are also older and more likely to own their own home.

14. An average of 17.8 teeth, more than twice as many. Dental health in America has improved. Thanks to fluoridated toothpaste and water and better dental care, baby boomers will do even better. But this has a downside. There will be twice as many teeth to be looked after in the year 2000 as there were ten years ago, but the number of dentists will peak a few years before that (at around 156,000) and then begin to decline. The population will keep growing. That means dentists will charge more, and many on fixed incomes may be unable to pay for the care.

15. The north edge of Alaska is at about 72 degrees north latitude. The west edge of the Aleutian Islands is at about 171 degrees east longitude. The south edge of Hawaii is at about 19 degrees north. And the east edge of Maine is at about 67 degrees east.

16. One in twenty-five families was headed by a single father in the early 1990s.

17. New Yorkers and Bostonians walk fastest.

18. Eleven million children see domestic violence each year.

19. Only 10 percent of absent fathers support their children. Their irresponsibility is a major cause of poverty in America.

20. About 80 percent of women will be widowed. They will live on their own for an average of sixteen years.

21. Because more women work outside the home. For every hour a woman works outside, she works a half hour less inside. Some men have tried to pick up the slack, though not very successfully. One study found that women spent an hour and a half less time cleaning house in 1985 than they had in 1965, and although men spent an hour more, that still left half an hour for things to fall apart.

22. Twenty-five percent of commuters breakfast in their cars.

23. Bananas. We eat 24.5 pounds of them per person every year.

24. The Bureau of Land Management controls about 274 million acres, 80 million acres more than the U.S. Forest Service. The National Park Service runs 80 million acres.

25. The Hispanic population, at current rates, will be larger than the black population in 2013, when there will be 42 million blacks and 42.1 million Hispanics. The main reasons are immigration and a high birthrate. In 1988, the Census Bureau estimated 100,000 illegal Hispanic immigrants would arrive each year for the next half century, but by the early 1990s, the estimate was doubled to 200,000 a year. Add in the legal immigration, and the Hispanic population will increase by 880,000 a year over the next half century. By the mid–twenty-first century, one in five of us will be Hispanic. Whites, with the lowest birthrate, will keep about the same number but drop from 75 percent of the population to about half. Overall, the non-white population will triple.

26. Alaska would win easily if it were a contiguous state, which it is not, being separated from the contiguous states by Canada. It has 33,904 miles of shoreline, but only 6,640 miles of coastline, which is how geographers refer to the general outline of the coast. Shoreline is the detailed measure of every bit of beach, including lakes, and the circumference of every island. By this measure Florida has the most shoreline: 8,426 miles of it. Louisiana is second with 7,721. Hawaii has only 1,052.

27. The Center on Hunger, Poverty and Nutrition Policy at Tufts University estimated in 1992 that some 30 million Americans are hungry. Others place the estimates lower, some as low as 5 million. The increase in poverty during the 1980s spread hunger more widely than before. Most hungry Americans are children. By 1992, 25.3 million Americans were receiving food stamps, an all-time high.

28. The average age in Llano County, Texas, is 49.4 years, making it home for the oldest in the nation. It also has the most residents over sixty-five (34 percent of 11,631 people).

29. The county with the youngest average age is Chattahoochee County, Georgia, where the average is 23.8 years old. But Wade Hampton County, Alaska, has the largest *proportion* of residents under eighteen: 45.8 percent of its population of 5,791.

30. There are some 700 drive-in theaters left in the United States, not counting the ones which have been transformed into flea markets and shopping centers. The invention and spread of television and the VCR, rising land values, changing sex mores—and smaller back seats?—did away with the old drive-in.

31. The United States was tenth worldwide in average income in 1991, according to the World Bank. Switzerland was first with $33,510 and Mozambique was last with $70. The United States placed behind Switzerland, Luxembourg, Japan, Sweden, Finland, Norway, Denmark, the former West Germany, and Iceland. This statistic, however, does not correct for what the money will buy in each country. With that correction added, the United States is number one.

32. The United States was twelfth in life expectancy. Japanese babies, which were first, can expect to live an average of seventy-nine years, while babies in Guinea-Bissau, which were last, are expected to live only thirty-nine years. Americans average seventy-six.

33. A study of black and Hispanic use of financial services found that about 36 percent of Hispanic-Americans have a major credit card. (The study excluded whites.)

34. About 35 percent of American blacks carry a major credit card.

35. Less than 5 percent of the artists represented are female.

36. About 85 percent of the nudes were women.

37. In the Library of Congress Geography and Map Reading Room, in the basement of the Madison Building, 101 Independence Avenue, S.E., Washington, D.C.

38. In a raised relief map, the surface is actually filled with bumps and hollows to show elevation; it is three-dimensional. A shaded relief map is two-dimensional with the relief effect gained by artistic shading. Among the finest examples of the latter are produced by Raven Maps and Images of Medford, Oregon, which offers a free color catalog. Phone 800/237-0798.

39. Half the lawyers in the world practice in America.

40. Trilingual, bilingual, American—so goes a wry joke.

41. About a third of American homes have computers. As we see ever-lower prices for ever-more-powerful computers, we also begin to see an astonishing computer glut in the nation's landfills. Ten million computers are discarded every year. If we continue at this rate, by the year 2005, there will be 150 million of them in the dump—enough to fill a hole one acre square and three and a half

miles deep. Disposal costs could run as high as a billion dollars. "It could be an enormous environmental problem," said Mark Greenwood of the Environmental Protection Agency.

42. Twenty-two percent of Americans, in 1993, felt it was possible the Holocaust never occurred. In the face of unquestionable evidence presented over a period of fifty years, this kind of voluntary ignorance deserves a phrase of its own. What's the opposite of blind faith? Blind skepticism?

43. The average American takes about 9,000 steps daily.

44. Depends on what you call America. Columbus never landed on any territory that is now a United State, though he did "discover," among other spots, the U.S. territories Puerto Rico and the Virgin Islands. Vikings, Phoenicians, and the Chinese are among others who historians have claimed discovered America before Columbus, but it was Columbus who opened the passageway for good.

45. St. Augustine, Florida, founded in 1565, is the longest-lived colony. The first attempted colonization of the United States was probably a Spanish settlement at San Miguel de Gualdape on the Georgia coast in 1526. It lasted about a year. The failed Roanoke Colony in Virginia (the fabled Lost Colony) was in 1585–90. "American" history has always been Anglophile, despite the Revolutionary War and War of 1812. Today's recognition of a multicultural America is beginning to rewrite the historical geography we teach in schools.

46. Nearly all the forty-eight contiguous states and Alaska are moving westward, generally, because of the continuing expansion of the Mid-Atlantic Ridge in the middle of the Atlantic Ocean. But there is a sliver of land attached to California that is headed northward, like a maverick heifer slipping away from the herd. This sliver, to the west of the San Andreas Fault, includes Los Angeles, which, if this movement continues, will be a suburb of San Francisco in about 20 million years. Hawaii, on the Pacific Plate, is heading northwestward.

47. A square mile, or a "section," has 640 acres.

48. Cape Alava, Washington, is the westernmost point.

49. Key West, Florida, is southernmost. In the entire United States the southernmost city is Ka Lae, Hawaii, 400 miles south of Key West.

50. West Quoddy Head, Maine, is the easternmost point.

51. The northernmost point is in the Northwest Angle of the Lake of the Woods, Minnesota, an extension of the forty-ninth parallel that is the apparent result of mapmaking errors beginning in 1783.

52. Niagara Falls, New York, had become a garish display of commercial establishments by the 1870s, and people realized they didn't want this to happen to the rest of the nation's scenic wonders. For Niagara Falls, as visitors can still see today, it was too late.

53. Colles published the first U.S. road maps.

54. The Oregon Trail, 2,000 miles from Independence, Missouri, crossing the Great Plains along the North Platte River, through the Rockies, through the high desert of southern Idaho, through the Blue Mountains of Eastern Oregon, along the Columbia River to the West Coast, has often been called the World's Longest Graveyard. The trail was originally a network of Indian paths and fur-trader trails. Between 1843 and the mid-1860s, some 350,000 people made the 160-day trip by horse, foot, and stagecoach, and one out of ten died along the way. The animals needed water, so the travelers followed rivers most of the distance, and many drowned.

55. A motel is defined as a place of lodging where you can drive right up to the door of your room. This wonderful idea, which eventually gave us the apotheosis of road lodging, the Holiday Inn (not to mention the Motel 6), came from a Los Angeles architect named Arthur Heineman, who created the first one in the world in San Luis Obispo, California, 200 miles north of Los Angeles along U.S. 101, in 1925. It had the perfect redundant name—Motel Inn—and it's still there.

56. They were all born in New Jersey.

57. Acadia has rugged seashore, Big Bend has desert mountains and river, Everglades has subtropical wetland, Mesa Verde has ancient Indian cliff dwellings, Sequoia has great trees and mountains, Zion has multicolored gorges, Yellowstone has geysers and hot springs, Olympic has midlatitude rain forest, Katmai has volcanic eruptions and big bears, Biscayne has coral reefs.

58. Salinas, California, now the home of one of your coauthors.

59. Illinois is twenty-fifth largest, Georgia is twenty-fourth, Iowa is twenty-sixth, and Delaware is forty-ninth.

60. The Louisiana Purchase, in 1803, almost doubled the size of the country by adding nearly everything between the Mississippi River and the Rocky Mountains.

61. Pogo lived in the Okefenokee Swamp on the Florida-Georgia border, place of "trembling waters" and home ground of such universal wisdom as, "We have met the enemy and he is us."

62. It's 5,859 miles across the country, measured from Elliott Key, Florida, to Kure Island, Hawaii.

63. Alaska.

64. Alaska.

65. Alaska—part of the state is across the 180-degree meridian and thus in the Eastern Hemisphere.

66. Ribbon Falls, in Yosemite National Park, has a single drop of 1,612 feet, second only to Venezuela's Angel Falls.

67. The United States and Canada, sharing language and history and politics, are members of the same culture realm (although there are Canadians who would deny this).

68. The U.S. Government, which provides information free or very inexpensively through the Consumer Information Center, Dept. 75, Pueblo, Colorado, 81009.

69. The farther south you go, the warmer the water, and oysters grow faster in warm water.

70. The United States has an average population density of 70.3 persons per square mile. What is it in your neighborhood?

71. Research suggests we may be a nation of neighbors—or at least something closer than strangers. Two out of three of us socialize with our neighbors. Half of Chicagoans spend a social evening with neighbors more than once a month. In Nashville, people knew an average of fifteen neighbors by name. People with higher education and income levels were slightly more neighborly. Blacks were more neighborly than whites. Women saw more of the neighbors than men, perhaps because it is often children who bring neighbors together and deepen their ties. "It's important to have neighbors as friends," a mother told researchers in the Bronx. "It's a good feeling to have a friend to count on in an emergency."

I

AMERICA'S NATURAL LANDSCAPES

The Lands

The lands we inhabit were brought into being by continuing interactions between the rocks below and the atmosphere above, the waters moving through both, and the activities of living things like beavers, termites, nematodes, and us. A single raindrop can create an instantaneous crater. Sand dunes and beaches change almost continuously. But it takes several million years for the Colorado River to erode the Grand Canyon of Arizona or for the dissolving of limestone to make a Carlsbad Caverns or Mammoth Cave, with their imagination-jogging, shape-shifting features.

Geographers call these things landforms, and their study is called geomorphology, or earth-shape science. We are all conscious of the shape of the land. We have to be. Otherwise we'd walk off cliffs, try to build parking lots on the tops of mountains, plant gardens in the rock. We are coming to understand more subtle truths as well. It might be wiser not to build houses on floodplains and in hurricane and earthquake zones, unless we're willing to pay the price in death, destruction, and insurance premiums.

1. What national park smells like Hell?

2. Why are the White Mountains called the White Mountains?

3. Why are the Green Mountains called the Green Mountains?

4. What do Mount Shasta, Mount Hood, Mount Rainier, and Mount St. Helens have in common?

5. Which has an older topography, the eastern United States or the West?

6. Where are the only mountains between the Appalachians and the Rockies?

7. Can you match the land use with the proportion of land allocated to each in America?

Rangeland	2 percent
Forestland	9 percent
Cropland	29 percent
Urban	30 percent
Protected	17 percent
Other	13 percent

8. If you've driven across northern Illinois and northern Iowa, you've crossed one of the country's smoothest, flattest landscapes, a "till plain" left by the last great Ice Age. What is a till plain, exactly?

9. After the great midwestern floods of 1993, some geographers were claiming that the levees built in the past century to prevent floods actually caused these floods. How could this be?

10. What do these have in common: the Green Mountains, the White Mountains, the Catskill Mountains, the Alleghenies, the Cumberlands, the Blue Ridge Mountains, and the Great Smokies?

11. Why are the Smoky Mountains called the Smoky Mountains?

12. Can you describe any of the four main ways in which mountains are formed?

13. Why, when you climb a mountain, getting closer to the sun with every step, does it get colder in-

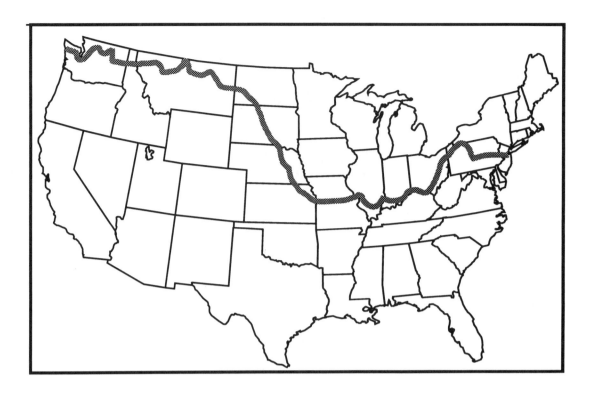

What does the line across this map of the forty-eight contiguous states represent?

Answer: The line marks the approximate southern boundary of the great continental ice sheets that crept across North America during the Pleistocene Ice Ages. The Pleistocene began about 2 million years ago and ended only 11,000 years ago. There were many Ice Ages, not just one. Each of the twelve to sixteen separate Ice Ages lasted for about 40,000 years. North of this line, the impact of the great ice sheets is still obvious in the soils and landforms. Indeed, in many places, the last 10,000 years have seen only slight changes.

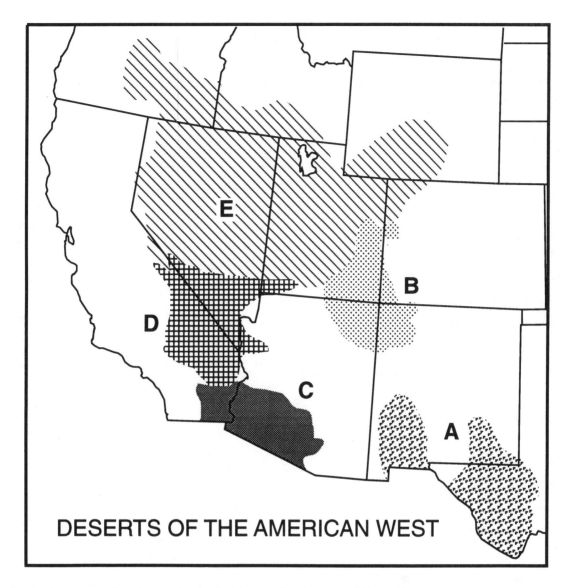

DESERTS OF THE AMERICAN WEST

Did you know there were five distinct kinds of desert in the American West? This map shows their location and relationship to one another. See if you can identify the deserts and match their letters with their names:

Sonoran Desert	E
Mojave Desert	B
Great Basin Desert	D
Chihuahuan Desert	C
Painted Desert	A

Answer: C is the Sonoran, D the Mojave, E the Great Basin, A the Chihuahuan, and B the Painted Desert.

stead of warmer? Sure, it has something to do with the thinning air—but what?

14. What two states share nearly all the Ozark Plateau?

15. Where is Two Ocean Pass?

16. Which of these sets of states are in the Piedmont?

New York, Massachusetts, and

Connecticut _____

Illinois, Iowa, and Indiana _____

Virginia, North Carolina, and

South Carolina _____

17. What is the most visited national park in the United States?

Yellowstone _____

Everglades _____

Yosemite _____

Great Smoky Mountains _____

18. Where was the Battle of Bunker Hill fought?

19. What are the Iron Ranges?

20. Which one of these states has the highest peak?

Colorado _____

Montana _____

Utah _____

Hawaii _____

21. What state has the highest low point? (Hint: Obviously, this will not be a coastal state where the lowest point is at sea level.)

22. Which state has the lowest high point?

23. The sixteen highest peaks in the United States are all in the same state. What state is that?

24. Why have the Rocky Mountains been worn down thousands of feet since they first came into existence?

25. A million years ago, ice sheets headed southward from Canada, finally extending into the Mississippi Valley and northern plains. Others—mountain glaciers—formed on mountains as far south as New Mexico. At their top speed, how fast would you say the ice sheets traveled?

An inch a day _____

A foot a day _____

A foot a week _____

A foot a month _____

A foot a year _____

An inch a year _____

26. What is, according to Theodore Roosevelt, "the one great sight which every American should see"?

27. What is being described here: first the rocky coast, then a mixed zone of rocky beach mixed with sandy beach, then an enormous coastline of mostly sandy beaches of crushed shell, followed by tiny coral islands, then a loop, an elbow of marshes and mud flats, and a final stretch of smooth, clean quartz-sand beach?

28. In New England, what feature is known as a "neck"?

29. For whom is Martha's Vineyard, an island off the Massachusetts coast, named?

30. In what state is Grand Island?

31. In what state is Grand Isle?

32. Where is Wizard Island?

33. What is a grain of sand?

34. Where in the United States can you find the largest concentration of seashells?

35. Can you name some places, besides the beach, where you will find sand dunes?

36. What are "sea stacks"?

37. Carlsbad Caverns, New Mexico, has a single room 350 feet high and many miles of galleries. Other well-known caves are Mammoth Cave in Kentucky, which has at least 150 miles of explored tunnels and passages as well as many more that have not been explored by anyone who came back to tell about it. How are such impressive underground features as caves formed?

38. Match these famous caves with their locations:

Mammoth Cave	Alabama
Carlsbad Caverns	Kentucky
Cave of the Wind	Oregon
Russell Cave	New Mexico
Oregon Caves	South Dakota
Luray Caverns	Virginia

39. What lives in Mammoth Cave?

40. Where is the biggest area of active volcanoes in, or even near, the United States?

41. What evidence of volcanic activity can you see from Manhattan Island?

42. Where are El Capitan, Half Dome, and Cathedral Rocks?

43. What are "badlands"?

44. In the Painted Desert in northeastern Arizona, you see plateaus in vivid reds, yellows, lavenders, vermilions, tans, beiges, khakis, and browns. What has caused these various colors to appear in the rock?

45. What is the Petrified Forest?

46. How is a "natural bridge" or arch of stone formed?

47. Match the desert with its most typical plant:

Great Basin	saguaro cactus
Mojave	Joshua tree
Chihuahuan	sagebrush
Sonoran	petrified wood
Painted	agave

48. What contains the following: two vast, shallow, shimmering salt beds, and the lowest spot on the continent, Badwater, elevation 279.6 feet below sea level.

49. What is the boundary between the geologically stable eastern United States and the shaky West?

50. Where is the thinnest part of the continental crust, and about how thick is it?

51. In colonial America, transportation along the coasts was primarily on water. Americans at that time thought of their coast as having four capes that had to be sailed around. What were these four capes?

52. Identify these famous islands:

Ellis Island _____

Angel Island _____

Gilligan's Island _____

Riker's Island _____

Three Mile Island _____

Treasure Island _____

Coney Island _____

Long Island _____

53. Are tides caused by the moon's gravity?

54. The most famous earthquake in U.S. history occurred on April 18, 1906. Where?

Answers

The Lands

1. Volcanoes National Park, on the big island of Hawaii, smells strongly of sulfur because of its main dramatic feature: volcanoes and craters such as Halemaumau—a name which sounds terrific when you say it aloud. The sulfur comes from inside the earth and is typical of volcanic eruptions. Sometimes in the park you can see heavy, thick waves of lava pushing their way across the main road.

2. The White Mountains are almost completely made up of pale gray granite.

3. The Green Mountains are covered with trees and other greenery, most notably ferns.

4. These are all great, individual peaks, and they are all part of the Cascade Range of mountains.

5. The eastern United States tends to have older topography. You can tell by the way the mountains are lower, more rounded and worn down by time. In the West, mountains are newer, more jagged, less restful looking. The Appalachians are more than 100 million years older than the Rockies.

6. The Boston and Ouachita mountains (the Ozarks) of western Arkansas are the only ones between the main eastern and western ranges. They were first inhabited 10,000 years ago by cliff dwellers.

7. The United States has 29 percent rangeland, 30 percent forestland, 17 percent cropland, 9 percent protected land. Two percent is urbanized and 13 percent other.

8. A till plain is a deposit of gravels scattered among clays and silts that got lodged under a glacier as it slid along. To see this happening today, you'd have to visit Greenland or Antarctica.

9. The floods were caused by truly great rains. However, levee construction, by not letting the waters flood naturally across the river valley floor, made the flooding far worse along those parts of the river where levees were never built, or were built too low.

10. These are all local names for different parts of the Appalachian Mountains, or Appalachian Highlands, in the East. The Rocky Mountains in the West, however, are not a single range, but a series of ranges: the Front Range in Colorado, the Grand Tetons in Wyoming, the Wasatch in Utah, and the Sawtooth in Idaho.

11. The Smoky Mountains are nearly always covered with a blue haze.

12. (a) The Sierra Nevadas were formed by a great fault, a clean break between neighboring portions of the earth's crust, one of which is shoved upward. (b) Some of the Rockies are domes, formed when molten rock, usually far below the surface, forced sedimentary rocks upward and then cooled. (c) Even more of the Rockies were formed when the earth's crust folded and humped, like a tablecloth when you push your plate away too hard. (d) Others are the remains of old volcanoes.

13. The atmosphere is heated primarily by reradiation from the earth's surface, not by solar energy entering the atmosphere for the first time. When the air is thinner, less dense, it absorbs less radiation from the sun, and less of the reradiation back

from the earth. You lose about three degrees of warmth for every 1,000 feet you climb. Another way to look at it is that every 1,000 feet you climb is about the same as, at sea level, driving 600 miles farther north.

14. The Ozark Plateau, now much eroded into steep hills and valleys, is shared by Missouri and Arkansas. Really picky geographers include a little of Oklahoma and an edge of Illinois as well.

15. Two Ocean Pass is a small lake in the Teton Mountains of Wyoming. It has two outlets. One is Atlantic Creek, which eventually finds its way into the Gulf of Mexico via the Mississippi River, and the other is Pacific Creek, which winds, via the Columbia River, to the Pacific.

16. Virginia, North Carolina, and South Carolina are in the Piedmont, a region of plains and low hills between the Atlantic Coastal Plain and the Appalachian Highlands.

17. Great Smoky Mountains is the most visited national park, probably because it is the most convenient to large population centers.

18. On Breed's Hill, across the Charles River from downtown Boston. In those days, June 17, 1775, the area was called Charles Town. Breed's Hill is what geomorphologists call a kame, what's left of an alluvial fan or delta that once formed on the front of a glacier.

19. These are six iron-bearing hilly ridges near Duluth, Minnesota. Iron ore from them was the basis for much of industrial expansion along the shores of the Great Lakes.

20. Puu Wekiu on the island of Hawaii is 13,796 feet above sea level, plus another 19,680 feet to its foot on the floor of the Atlantic, making it higher than any part of the other "mountain" states.

21. Colorado has the highest mean elevation of any state—6,800 feet—and its lowest point (in the channel of the Arkansas River at the border with Kansas) is 3,350 feet above sea level.

22. Delaware has the lowest average elevation of any state—about 60 feet above sea level. But Florida's highest point is the lowest: only 345 feet above sea level.

23. Alaska.

24. The Rockies, like other mountains, begin wearing down through erosion as soon as they are formed. Rain and snow-melt pour in torrents down the mountain flanks, carrying boulders, stones, and soil along with them, and spreading these over the Great Plains.

25. The ice sheets sometimes barreled along at a foot a day. As they went, they altered the land in two ways: by bringing, and by taking away. They brought topsoil from the New England mountains, for instance, leaving them largely barren, and deposited it in what is now Connecticut, enriching that state and endowing it with fertile fields full of rocks. Give a farmer rocks, he makes a wall. You see them all over New England.

26. The Grand Canyon, a magnificent ditch cut by the Colorado River, second-longest in the United States. The canyon itself is 220 miles long.

27. This is the East Coast of the United States followed by the Gulf Coast.

28. A neck is a sandbar connecting an island—the "head"—to the mainland.

29. Martha's Vineyard is named for the daughter of Bartholomew Gosnold, an English explorer who cruised the region in 1602.

30. Nebraska is home of Grand Island.

31. Louisiana is home of Grand Isle.

32. Wizard Island is a volcanic cone inside Crater Lake, in Oregon.

33. A grain of sand is a minute bit of weathered rock—usually quartz, but often feldspar or limestone or something else—somewhere between two-thousandths and eighty-thousandths of an inch in diameter. Most end up on beaches, having originated hundreds of miles away on the tops of mountains where a particular grain may have begun its career as a boulder. Sand grains, once they reach their ideal size, tend to survive from then on, encased in a protective skin of water: Grains of wet sand often slip past each other without wearing. Dry grains higher up on the beach are still being ground down.

34. On Sanibel and Captiva islands, off the west coast of Florida, because of a rare combination of currents and tides. These islands are settled by humans now, and frequently visited by hordes of "shellers"—tourists who come to collect the mollusks' husks. These are merely calcium carbonate, secreted, colored, smoothed, and pearlized by the animals' mantle. Shellers have not managed to completely clear the beaches of shells, but they—and development—have nearly leveled shell mounds, some of them 50 feet high, that were left after oyster lunches by the early Indians.

35. Great Dunes National Monument in Colorado has dunes 600 feet high, and White Sands, New Mexico, has 500 square miles of gypsum dunes (gypsum is the mineral from which we make plaster of Paris). Hills all around the Great Lakes are ancient dunes, now covered with vegetation. Dunes are still being formed around the Great Lakes, as they are anyplace where you find sand and wild winds.

36. Sea stacks are the chimneylike remnants of vertical sea cliffs and headlands cut off from the mainland. They are a dramatic feature of many parts of the Pacific coast. Sea stacks are created by the sea, which during a powerful storm can slam into the rocks with the force of many tons per square foot.

37. Today's cave started out as a vast deposit of limestone on the bottom of the sea. Shifts in the earth's crust lifted it up, and over the centuries and millennia seeping water dissolved the limestone because it was softer than the surrounding rock. The slowly forming hollows became caves. If the dripping continues, the deposited solution forms stalagmites and stalactites.

38. Mammoth Cave is in Kentucky, Carlsbad Caverns in New Mexico, Cave of the Wind in South Dakota, Russell Cave in Alabama, Oregon Caves in Oregon, Luray Caverns in Virginia.

39. Two-inch-long colorless fish called blindfish. Though sightless, these creatures seem to thrive in the cave's dark, using sensitive nerves in their heads to detect movement in the water. Presumably something else lives there and stirs in the water and provides the blindfish with food. It would seem obvious that they became blind because they live in a lightless environment, but a more recent theory postulates they were blind when they lived above ground, found it difficult to compete, and then luckily blundered into the cave where, in the land of darkness, the blindfish is king.

40. The Valley of Ten Thousand Smokes at the base of the Aleutian Island chain in Alaska is the biggest area of active volcanoes.

41. The Palisades, across the Hudson River in New Jersey, are a ridge of volcanic rock. They formed when the North American continent drifted away from Africa more than 100 million years ago.

42. In Yosemite Valley.

43. A badland is an area in the plains where erosion has carried away the soft, sedimentary rock so that only residual pinnacles are left. The result is a region of slopes eroding so quickly they have no soil or vegetation, even if the climate is wet. Usually, a badland started life as a high plateau that eroded as a river system clawed downward into it. Badlands National Monument in South Dakota is the most bizarre and dramatic example.

44. Rusting metal, in a sense. Weathering painted the Painted Desert. Various compounds within the rock, mostly common variants of iron, change color as they oxidize upon exposure to air, sun, and water for different amounts of time.

45. Also in the Painted Desert is the Petrified Forest—the trunks of trees which grew 150 million years ago and, upon falling, were buried in shallow lakes and marshes where air could not reach.

Ground water leached away the wood fibers and replaced them, molecule by molecule, with minerals like silica, leaving behind "trees" of glassy stone. They lay 3,000 feet beneath the surface until they were unearthed by erosion.

46. A river is flowing along in a deep stone bed. At some point, it makes a hairpin turn, a kind of horseshoe-shaped loop. But the river wants to go straight and keeps pounding at the neck of the loop until it breaks through. If the channel is deep enough, an arch or bridge is left overhead. The most dramatic examples may be at Bridges National Monument and Arches National Monument in Utah.

47. The Great Basin Desert is most typically characterized by sagebrush, the Mojave by its grotesque Joshua tree, the Chihuahuan by agave, the Sonoran by saguaro cactus, and the Painted by petrified wood.

48. Death Valley, certainly a low dive. Curiously, only 80 miles away is the highest spot on the U.S. mainland, Mount Whitney, which is 14,495 feet high. Death Valley gets an average annual rainfall of about two inches. The name Death Valley is a mere poetic metaphor—one study found 608 different kinds of plants flourishing within the valley's confines.

49. The Rocky Mountains.

50. The thinnest is in the region geographers call the Basin and Range, between the Rocky Mountains of Colorado and the Sierra Nevadas of California. Here, the earth is stretching apart for reasons that are unclear, and the continental crust is only 20 miles thick.

51. Cape Cod in Massachusetts; most of New Jersey between the Delaware and Hudson River estuaries, the outer point being Cape May; the Delmarva Peninsula between the Delaware and Chesapeake estuaries; and Cape Hatteras in North Carolina.

52. Ellis Island was the clearing center for immigrants in New York harbor; Angel Island was the clearing center for immigrants in San Francisco Bay; *Gilligan's Island* was a TV show starring Bob Denver; Riker's Island is a penitentiary on an island in the East River off the Bronx; Three Mile Island is the site of a nuclear power plant in Pennsylvania. There are at least two Treasure Islands, in addition to Robert Louis Stevenson's fictional one: on the west coast of Florida, near St. Petersburg, and adjacent to Yerba Buena Island in the middle of the San Francisco-Oakland Bay Bridge. Coney Island is in New York, site of a famous beach resort and amusement park, where the natives are said, by the writer Ralph Lombreglia, to have 100 words for Sno-Kones, and Long Island is a long island east of Manhattan.

53. By the moon's gravity, and also by that of the sun and other planets. The highest tides come twice a month—when the moon is full and when it's new. This occurs when Earth, moon, and sun are in a straight line. These are called spring tides. During the quarter moon, also twice a month, the tides—then called neap tides—are at their lowest. Spring tides have the highest range—highest high and lowest low. Neap tides have the lowest range—from lowest high and highest low. You may have noticed that the tides of each day are fifty minutes behind the tides of the day before, just as the moon rises fifty minutes later each night.

54. This was the infamous San Francisco Earthquake, whose movement occurred along the San Andreas fault.

The Waters

Water—salt or fresh, still or running, frozen or liquid or vapor—is the most immediate link we have to the natural environment. It is central to our lives. If you show photos of different scenes to people anywhere in the world and ask them which place they prefer (as geographers have done), they'll always pick the one with water in it. If you add water to a scene that ranked far down in the list without water, the scene leaps toward the head of the list. We react similarly to images of vegetation, but much less powerfully.

Nearly every important archaeological site is on water of some kind, typically a river, stream, spring site, or shoreline. We use it for recreation, navigation, transportation, irrigation, cleaning, hydroelectricity production, flushing toilets.

Or we simply gaze at it happily, like subjects in a geographer's experiment.

1. There are five states along the western bank of the Mississippi River. How many can you name? What are their capitals?

2. "The _____ and the _____ have made a deeper impression on me than any other part of the world." Fill in the two proper names in this quotation from T. S. Eliot.

3. What causes "inland water"—rivers, lakes, creeks, etc.?

4. The United States has the shortest named river in the world. Where is it?

5. Match these rivers with the state of their source:

Mississippi	New Mexico
Ohio	North Carolina
Missouri	Pennsylvania
Tennessee	Minnesota
Atchafalaya (via Red River)	Montana

6. Where is the highest dam in the United States and what river does it block?

7. The ultimate headwaters of the St. Lawrence River are in which state?

8. What is America's most important "exotic" river? (Hint: An exotic river is one that flows from a humid environment into an arid environment.)

9. Rivers form patterns, each for its own good reasons. "Dendritic" or "parallel" rivers flow toward the sea on flat land with a gentle slope. "Trellis" rivers flow where the soil or rock is unevenly resistant. Where and why would there be a "radial" river pattern?

10. How far north of the Gulf of Mexico is the source of the Mississippi River?

11. What state was created in its virtual entirety by the Mississippi River?

12. What Nebraska river was an important route for the westward pioneers?

13. What Nevada river was an important route for the pioneers traveling to California?

14. Only one river cuts all the way through the Appalachian Mountains. What is it?

15. What bridge connects Brooklyn and Manhattan?

16. Where is the longest suspension bridge in the United States?

17. What bridge connects San Francisco and Marin County?

Match these eastern rivers with their letters:

Ohio	Chattahoochee	Wabash
Mississippi	Cumberland	Illinois
Savannah	James	Tennessee
Connecticut	Susquehanna	Hudson
Black Warrior	Wisconsin	

Answers: A is the Mississippi, C the Black Warrior, E the Tennessee, D the Chatta-hoochee, B the Cumberland, F the Ohio, G the Illinois, M the Wabash, H the Wis-consin, J the Hudson, K the Susquehanna, L the Connecticut, N the James, P the Savannah.

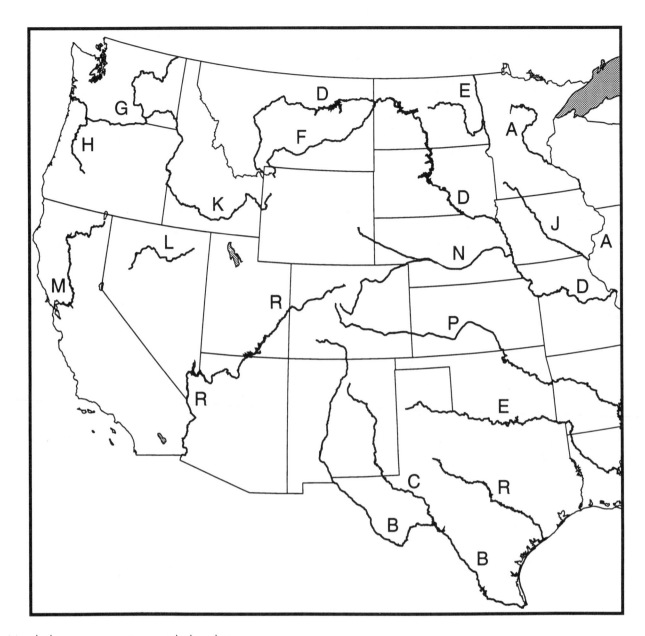

Match these western rivers with their letters:

Humboldt	Platte	Colorado
Mississippi	Columbia	Pecos
Rio Grande	Willamette	Snake
Sacramento	Missouri	Iowa
Yellowstone	Arkansas	Red

Answers: A is the Mississippi, B the Rio Grande, C the Pecos, R the Colorado, P the Arkansas, E the Red, D the Missouri, N the Platte, F the Yellowstone, H the Willamette, G the Columbia, K the Snake, J the Iowa, L the Humboldt, M the Sacramento. (Yes, there are two Red Rivers and two Colorado Rivers.)

18. What two peninsulas does the Mackinac Bridge connect?

19. The Julia Tuttle Causeway connects what two cities?

20. What watery feature, built between 1811 and 1825, was 4 feet deep, 40 feet wide, and had paths on each side for mules?

21. Where is the world's largest dam, as measured by volume?

22. Which Great Lake is the highest?

23. Besides fish and water skiers, what do Lake Havasu, Lake Mojave, Lake Mead, and Lake Powell all have in common?

24. Lakes never remain the same. Describe the way in which they tend to change.

25. Where are the most lake-filled parts of the United States?

26. Where was the biggest lake in the United States? (Hint: It doesn't exist as a lake anymore.)

27. What happens to lakes surrounded closely by vegetation?

28. What keeps lakes filled and rivers running even during drought periods?

29. How large was Lake Bonneville?

30. Lake Bonneville left behind Great Salt Lake, Little Salt Lake, Lake Sevier, and Lake Utah. All are salt lakes except one. Can you name that one and tell why it is fresh?

31. The Great Salt Lake, the largest saltwater lake in the Western Hemisphere, is salty to the tongue, but noisome to the nose. Why?

32. Do the western deserts contain other kinds of lakes besides freshwater lakes and saltwater lakes? If so, what kinds?

33. In California, largely a desert state and the country's most populous, a single use consumes by far most of the state's water. What is it?

34. What is the Ogalala Aquifer?

35. What is swampbusting?

36. In what direction is the spread of irrigated agriculture moving in the United States?

37. What state uses the most water per capita? (Hint: Think of a state with lots of irrigated agriculture and a low population.)

38. What was the worst marine disaster in U.S. history?

39. Where in the United States would you find the most hot springs, geysers, hot pools, natural kettles of boiling mud, and fumaroles—all in one national park?

40. What makes a geyser?

41. Why is Old Faithful sitting atop a 12-foot-wide, 55-foot-high mound?

42. What causes flashfloods in the Southwest?

43. "More than 80 percent of the earth's v_____ e_____ are thought to occur in the oceans." Add the two missing words to this sentence from a *New York Times* story about a cataclysmic event no one saw.

44. The four highest waterfalls in the United States are all found in a single national park. Can you name it?

45. America is getting lower. According to one estimate, erosion by water is lowering the entire surface of the continent. Guess by how much?

About an inch a year _____

About a foot a year _____

About a foot and a half a year _____

About a foot in 10 years _____

About a foot in 100 years _____

About a foot in 1,000 years _____

About a foot in 10,000 years _____

46. What creates a waterfall?

47. Do waterfalls last forever?

48. The greatest waterfalls in America—some 3,000 of them—have one thing in common. What is it?

49. According to tagging studies of salmon, what proportion of them manage to find the same stream where they were born?

50. What creature's life story is a kind of mirror image of the salmon's?

51. Look down at any map of the United States and notice, around the Gulf of Mexico, the places where the shoreline sort of bulges out. What do you almost always see just inland?

52. Why don't you find coral reefs anywhere in the United States except off southern Florida?

53. Which direction does beach sand move on both the east and west coasts of the United States?

54. Where did the original Hawaiian Islanders come from?

55. What is the only major nonfederal waterway in the United States?

56. Where is the tallest water tower in the world?

The Waters

1. The five states on the west bank of the Mississippi and their capitals are, from the south: Baton Rouge, Louisiana; Little Rock, Arkansas; Jefferson City, Missouri; Des Moines, Iowa; and St. Paul, Minnesota.

2. The Missouri and the Mississippi rivers made quite an impression on Eliot.

3. Inland water is simply the result of more rain falling on the continent than can be absorbed by the soil or used up by plants. It lies there in depressions called lakes or ponds, or runs downhill toward the sea in rivers. It also, over time, alters the surface of the earth, creating such wonders as the Grand Canyon. Water moving at only a foot per second can carry gravel with it, and gravel can scour holes in the landscape. Water gurgling along at 25 feet per second can roll boulders.

4. According to the *Guinness Book of World Records,* the shortest river is the mighty Roe River, a 201-foot tributary of the Missouri River in Montana.

5. The Mississippi originates in Minnesota, the Ohio in Pennsylvania, the Missouri in Montana, the Tennessee in North Carolina, and the Atchafalaya in New Mexico. Yes, New Mexico.

6. The Oroville Dam on the Feather River in northern California is highest, at 770 feet. Hoover Dam is only 726 feet high.

7. Minnesota, via the North River, is the ultimate headwaters of the St. Lawrence.

8. The Colorado River, beginning in the Rocky Mountains and flowing into the Gulf of California, is the country's most conspicuous exotic river.

9. The radial pattern occurs when rivers flow down all sides of a mountain, creating a spokelike design with the top of the mountain as the hub.

10. The Mississippi gets its start as an unprepossessing creek flowing from Lake Itasca, in Minnesota, 2,500 miles north of the Gulf.

11. Louisiana was largely deposited by the Mississippi. It is made of bits of soil from other states farther north.

12. The Platte.

13. The Humboldt.

14. The New River of West Virginia, which connects to the Roanoke River of Virginia downstream, cuts all the way through the Appalachians. Rough and narrow, it never became highly important for navigation.

15. The Brooklyn Bridge, spanning the East River, connects Brooklyn and Manhattan.

16. The longest suspension bridge is the Verrazano Narrows Bridge near New York City. It is 60 feet longer than California's Golden Gate Bridge.

17. The Golden Gate Bridge connects San Francisco and Marin County.

18. The Upper and Lower peninsulas of Michigan are connected by the Mackinac Bridge.

19. Miami and Miami Beach are connected by the Julia Tuttle Causeway.

20. The Erie Canal, which was constructed inland from Albany, New York, and eventually became

363 miles long, each section having been opened as it was completed.

21. The largest dam in the world is in Arizona. It is called the New Cornelia Tailings Dam, and it dams not water, but the waste from a copper mine.

22. Lake Superior is the highest Great Lake, at 600 feet above sea level.

23. All are behind large dams (Parker, Davis, Hoover, and Glen Canyon, respectively) on the Colorado River.

24. Wherever a lake appears, either as a crater atop a mountain that once was a volcano, or as an ox-bow—the cut-off elbow of a sinuous river—it immediately begins changing. Streams flow into it, depositing sediment. Vegetation creeps in. Streams flow out of it, cutting their own channels deeper and letting the lake drain lower. Lakes always get smaller, and shallower, and in the end become dry land.

25. Alaska, New England, Minnesota, and Wisconsin are the lakiest parts of the country, mainly because of their history of glacier occupation. Glaciers traveled down mountain valleys and made finger lakes, such as those in western New York and in Glacier Park in Montana. They left behind great blocks of ice that later melted and formed lakes. Minnesota is the "Land of 10,000 Lakes," although it is not clear that anyone has actually counted, and if they have, whether the number might not, by now, be 9,999—lakes being the impermanent geographic features they are.

26. Lake Agassiz, as geologists call it, once covered western Minnesota and eastern North Dakota, extending far northward into Saskatchewan and Manitoba. At its largest, it was 250 miles wide and 700 miles long, greater than all the Great Lakes combined. All that remains are Lake of the Woods, Lake Winnipeg, and vast, fertile wheatlands.

27. Lakes surrounded closely by vegetation disappear even more quickly than others. The plants send roots along under the surface, holding silt and sediment, slowly filling and strangling the lake and turning it into a bog.

28. Underground water—or the groundwater—keeps lakes filled and rivers running during drought periods, and accounts for the vast majority of the water in America. It tends to sink slowly downward through the seams of the earth, but when it encounters something impermeable, such as hard-packed clay or a stone shelf, it moves horizontally along these features until it, like all other moving water, finally reaches the sea. If it finds a good spot to emerge, it may pop up as a spring. Or a geyser.

29. Lake Bonneville, which has shrunk and dried down to a few remnant lakes, such as the Great Salt Lake, once was 1,000 feet deeper than the Great Salt Lake. A gigantic inland lake of glacier melt, it covered parts of southern Idaho, western Utah, and eastern Nevada.

30. Lake Utah is fed by mountain streams and has an outlet (into Great Salt Lake), so it has not, like the others through years of repeated evaporation and concentration, become saltier and saltier.

31. Sometimes, the Great Salt Lake is seven times saltier than the ocean. At its saltiest, during a dry season, it is saltier than the Dead Sea. In fact, Great Salt is so thick with salt that diving beneath its surface is nearly impossible (no one has ever drowned there). Great Salt is also a bitter lake, containing sulfates. But Great Salt is home to billions of brine shrimp, and what looks like sand on the shore is actually the hardened pellets of their excrement. Brine flies swarm along the beach, and so do millions of migrating birds from as far away as Tierra del Fuego. The birds pursue the shrimp and leave their own excrement behind. The result, as Congresswoman Karen Shepherd of Salt Lake City puts it, is that "In August, you can't get near it because it smells so bad."

32. Yes, there are also bitter lakes, which contain sulfates; and borax lakes, which contain borate minerals.

33. Irrigation consumes nearly 90 percent of California's water. In the more humid East, industry consumes most of the water.

34. The Ogalala Aquifer is the water-saturated layer of sand and gravel deep below much of the Great Plains from Texas to Nebraska. Most of this water is what is called fossil water, that is, it has been there a long time without being replenished by modern rains. In recent years, wells have begun to tap so much of this aquifer that it is beginning to show signs of drying up.

35. If "sodbusting" is plowing up and developing the American prairies, then "swampbusting" is draining and filling the wetlands. Wetlands are crucial to wildlife, and in some areas swampbusting is doing irreversible damage.

36. Irrigated agriculture is moving eastward in the United States. It protects against the effects of even brief summer droughts and allows better use of rich midwestern soils. Irrigation has increased especially in the Mississippi Valley.

37. Idaho uses about 22,000 gallons a day per person.

38. Although 1,513 persons went down with the *Titanic* in 1912, the worst marine disaster took at least 1,547 lives when the Mississippi sidewheeler *Sultana* blew up on April 27, 1865, about eight miles out of Memphis. Most of the dead were Union soldiers returning home from Confederate prison camps.

39. At Yellowstone National Park, which has at least 10,000 of them. Fumaroles are vents of steam. Old Faithful is, of course, our star geyser.

40. A geyser needs three things: underground water; a hollow pipe of rock to carry it to the surface; and, deep underneath, a seething pool of molten rock, or magma, that will superheat the water, turning some of it into steam and causing the steam and water to jet from the pipe of stone. A hot spring, such as those at Hot Springs, Arkansas, is precisely the same thing minus steam pressure.

41. Old Faithful, though a young geyser as geysers go, has carried to the surface with its steamy water—about 12,000 gallons per squirt, a squirt every sixty-four and a half minutes on the average, for about the past 300 years—loads of dissolved minerals and deposited them around itself to form this gigantic mound. Exactly the same thing happens inside the teakettle of those who live in hard-water parts of the country.

42. Desert streams usually flow in rock-bottomed basins covered by thin soils and little vegetation. When a rare heavy rainfall occurs, the water usually is not absorbed quickly enough to prevent flash flooding. As we cover more and more landscape with parking lots and roofs, we are making even humid regions behave more like deserts. As cities expand, such flashfloods are an increasing possibility.

43. Volcanic eruptions.

44. All four of the nation's highest waterfalls are in Yosemite National Park, California.

45. America is lowering about one foot in 10,000 years.

46. A waterfall is created by water flowing over a weaker portion of a riverbed. It cuts deeper and deeper until finally it is falling into a gorge from a great height.

47. The roiling water of a waterfall undercuts its own lip, creating those dim, roaring, misty, romantic caverns you see in novels and movies—you know, where the bickering couple we knew were attracted to each other first acknowledge their love? Then the lip collapses and the waterfall has moved upstream, which it continues to do until it is no more. Niagara Falls is retreating upstream about 4 feet a year, and has moved 7 miles since it was born.

48. They are manmade—they are dams. The Grand Coulee Dam on the Columbia River, for instance, is four times as high as Niagara, with 117,000 cubic feet of water pouring over the top of it *every second.*

49. Not all salmon find their way home, but about 90 percent do, plowing mindlessly yet somehow brilliantly upstream at an average of 6 miles per day. Some have been observed leaping up 10-foot waterfalls.

50. While the salmon is born in her river, dwells most of her life in the sea, then returns to her home river, the Atlantic eel is born in the ocean, lives his life span in rivers and lakes, and then returns to his ocean to spawn.

51. A river is almost always inland of the bulges. These are delta regions. The Apalachicola, Chattahoochee, Mississippi, Brazos, and Rio Grande have helped form the Gulf Coast, bringing soil from the country's interior down to this all-but-enclosed sea and building up the land.

52. The coral animals of which the reefs are composed grow only in warm, shallow seas—with temperatures above 68 degrees Fahrenheit—where plenty of light penetrates the once-clear water, and currents bring a flow of food. Most U.S. coral reefs are seriously threatened by sewage and pesticide runoff and ignorant boaters and divers.

53. On both coasts, beach sand moves predominantly south. The reason is that waves arriving at both coasts start in storms in the northern parts of the Atlantic and Pacific oceans. These waves then wash down from the northeast and northwest, respectively, causing the sands in the surf zone to move southward. On both coasts, you can see the result as sands build up on the north sides of jetties and groins.

54. The best speculation is that Hawaiians came from the Marquesas Islands, far to the south, in about the third century A.D. They were Polynesians, similar to the original peoples of Tahiti and the Maori of New Zealand.

55. The 522-mile-long New York State Barge Canal System.

56. Union, New Jersey, has the tallest water tower. At 210 feet, it holds 250,000 gallons.

Weather

In Camelot it rained only at night. How convenient. We seem to become alarmed and resentful when the weather fails to fulfill our expectations. But of course, that isn't weather's job. Weather is *supposed* to be variable—if it weren't, we'd have to call it something else.

Weather is the daily goings-on in the atmosphere above us—what it does. Climate, on the other hand, is the long-term average condition of the atmosphere above a place—what it's like.

To some extent, weather is predictable. It is a fact of world climate, for instance, that high pressure brings fine, dry weather, while low pressure is likely to bring rain—or something more unpleasant, such as Hurricane Andrew, which knocked down big trees onto the house of one of your coauthors in August of 1992. The United States, particularly when you include Hawaii and Alaska, has more types of climate and weather than any other country in the world.

Both are always there, a part of daily life, influencing us in ways we often don't even know. The psychological effects of Hurricane Andrew, experts found, persisted more than a year after the storm blew over.

1. How did Dorothy and her dog Toto travel to Oz?

2. What would you say is the average temperature of the world?

    33 degrees Fahrenheit _____

    47 degrees Fahrenheit _____

    59 degrees Fahrenheit _____

    67 degrees Fahrenheit _____

3. In 1816, something very unusual occurred in New England weather. A June snow blanketed much of the region and the entire summer was very cold. What caused this ''year without summer,'' as it was called?

4. Ten of these are the ten driest cities in the United States, and the eleventh is not. Pick the one that is not:

    Yuma _____

    Phoenix _____

    Winslow _____

    Las Vegas _____

    Reno _____

    Winnemucca _____

    Bishop _____

    Bakersfield _____

    San Diego _____

    Alamosa _____

    El Paso _____

5. The 100th meridian passes from north to south across the western Dakotas to Texas. What does it divide?

6. Not many years ago, the highest wind velocity ever documented on the planet was recorded, not in a hurricane or tornado, but in a place that would probably be impervious to either of those monsters of weather if one struck. Can you guess where?

7. What layer of the atmosphere do we live in?

    The troposphere _____

    The stratosphere _____

    The stradivariusphere _____

The pherosphere _____

The pollutosphere _____

The shakesphere _____

8. Beside each of these states is a number representing the number of _____ s it had per thousand square miles between 1953 and 1991. Fill in the blank.

Florida   30

Oklahoma   29

Tennessee   23

Vermont   3

Maine   2

Alaska   0

9. "In the spring I have counted one hundred and thirty-six different kinds of weather inside of twenty-four hours." Of what state was Mark Twain speaking?

10. What U.S. city has the highest average temperature?

Miami _____

Key West _____

San Diego _____

Chula Vista _____

Needles _____

Tucson _____

Honolulu _____

San Juan _____

11. The hottest day ever recorded in the United States was, you probably figured, in Death Valley, California. Can you guess the temperature within 10 degrees?

12. The coldest place ever recorded in the continental United States was Rogers Pass, Montana, on January 20, 1954. Guess the temperature within 10 degrees.

13. How many inches of rain would you say fell on Alvin, Texas, on July 25–26, 1979, producing the record for most rainfall in a twenty-four-hour period?

14. And the record for snow?

15. What is the hottest temperature ever recorded at a weather station in Alaska?

16. What is the coldest temperature ever recorded at a weather station in Hawaii?

17. What is the coldest temperature ever recorded at a weather station in Florida?

18. The U.S. Weather Service has a "Discomfort Index" based on temperature and humidity. What U.S. cities would probably rate highest in discomfort?

19. What is the climatic difference between the inner, built-up part of a city, and a green park in the same city?

20. What is the difference between a tornado and a dust devil?

21. In what region of the United States do tornadoes almost never occur?

22. What is America's most common natural disaster?

Fire _____

Flood _____

Hurricane _____

Tornado _____

Earthquake _____

23. The United States has ten different climate types, the greatest variety of any nation. How many can you name or describe?

24. What is a "Nor'easter"?

25. What was the worst natural disaster in United States history?

26. What is vog?

27. Where in the United States do trade winds blow?

28. What part of the United States is south of the Tropic of Cancer, and what is the Tropic of Cancer?

29. Why is it so unpleasantly hot and humid in the summer in the Midwest?

30. Who were the Weathermen, and what was the Weather Underground?

Weather

1. Dorothy and her dog left Kansas, house and all, in a tornado.

2. The earthly average temperature is 59 degrees Fahrenheit.

3. Today's explanation is that a volcanic eruption in Indonesia the year before spread so much ash into the atmosphere that a significant portion of the sun's energy was reflected back to space instead of reaching, and warming, New England.

4. San Diego is the odd city out.

5. The 100th meridian is the general dividing line between the humid East and the mostly arid West.

6. The highest wind, 231 miles per hour, was recorded on the exposed summit of 6,290-foot Mount Washington in New England, the highest point north of the Smokies and east of the Black Hills. Here, snow falls in every month of the year.

7. We live in the troposphere, the lowest layer of atmosphere, usually about ten miles thick. Above it is the stratosphere, where the ozone holes are. If you ever get to the stradivariusphere, you will hear faint, ethereal fiddle music.

8. The number represents the number of tornadoes per thousand square miles each state sustained. Florida had the most; Alaska, South Carolina, Nevada, and Oregon had none. Tornadoes are among the most destructive weather events of all. Their wind power is astonishing. Witnesses saw two cows in the sky during a bad one in 1973 in Union City, Oklahoma. One came to earth and lived, the other was never seen again.

9. Connecticut.

10. Miami's average temperature is 75.6 degrees, Honolulu's is 77 degrees, but San Juan, Puerto Rico's is 79.7 degrees. The average of very hot desert cities is reduced by their cool night temperatures.

11. The hottest day was July 10, 1913, and the temperature was 134 degrees. The measurement was taken, incidentally, at a place called Greenland Ranch, which must be a misnomer.

12. The record cold temperature was minus 70 degrees Fahrenheit.

13. Alvin got 43 inches of precipitation.

14. Silver Lake, Colorado, got 76 inches of snow on April 14–15, 1921. The most snow in a year fell at Paradise Ranger Station, Washington, in 1971–72, which got 1,122 inches—nearly 100 feet.

15. Fort Yukon had 100 degrees Fahrenheit on June 27, 1905.

16. It was minus 17 degrees Fahrenheit at Haleakala, Maui, on January 2, 1961—at an elevation of 9,750 feet.

17. It was minus 2 in Tallahassee on February 13, 1899—at an elevation of 193 feet.

18. Houston and Phoenix are rated the two most miserable cities, with a Discomfort Index of 85. They are followed by Key West, with 84, and Miami-Fort Lauderdale with 83.

19. About 8 to 10 degrees, according to city planners. The concrete retains and radiates heat.

20. A dust devil is a whirling mass of air created, only in the daytime, by the heating effect of the sun. It starts on the ground. A tornado starts in the atmosphere and may occur in day or night. Tornadoes often don't touch the ground, but if one does, that is when it is most dangerous, sometimes devastating any weather instruments it hits.

21. Tornadoes seldom, if ever, occur west of the Rocky Mountains.

22. Flood is the most common natural disaster by far, accounting for more than 90 percent of presidential disaster declarations. Hurricanes and tornadoes, terrible as they can be, strike and move on, while floodwaters remain behind and continue to do damage. Nationwide, nearly 8 million homes are exposed to the threat of floods. Only 2 million have flood insurance.

23. The Northeast has a humid continental climate, with warm, humid summers, cold, snowy winters, and four distinct seasons. The Southeast has a humid subtropical climate, with hot, humid summers and mild winters. The tip of Florida has a humid tropical—with winter drought—climate, warm throughout the year. The Great Plains have a continental steppe climate, semiarid and supporting grasslands but not forests, with hot summers and very cold winters. The interior western states, dominated by mountains, have a varied highlands climate, and desert and steppe climates in between. On the Pacific Coast we have two climates, the maritime in the north, mild and moist; and the Mediterranean in southern and central California, with long, dry, sunny summers and mild, somewhat wetter winters. Hawaii has a humid tropical climate. Alaska has a maritime climate in the south, a

subarctic climate in the interior, and a tundra climate area in the far north.

24. A "Nor'easter," or Northeaster, common in Maine, is a winter storm off the Atlantic which brings strong winds, large ocean waves, and cold, snowy weather.

25. Many consider Hurricane Andrew, of August 24, 1992, which ripped across South Florida, to have been the worst. By one estimate, it destroyed 100,000 houses and left 250,000 homeless at a potential cost of at least $20 billion. A year later, 27 percent of the people of South Dade County, where the center of the storm hit, had moved away. Only a handful were killed because Andrew, though powerful, passed across quickly and was a relatively dry hurricane. Most hurricane deaths are by drowning.

26. Vog is a new Hawaiian word for a fog made up of volcanic ash.

27. Trade winds blow from subtropical high pressure cells, which are usually over oceans in the tropical latitudes toward the Equator. The only two parts of the United States regularly swept by the trades are Hawaii and South Florida—plus Puerto Rico and the U.S. Virgin Islands, of course.

28. Hawaii. The Tropic of Cancer is the northernmost latitude where the sun is vertical at noon on at least one day of the year. It is 23.5 degrees north latitude and Hawaii is mostly between 20 and 22 degrees. Texas and South Florida are too far north.

29. The Midwest is hot and humid in summer because of "mT." These are the code letters geographers use to describe regional air masses that sweep across the country. The "m" stands for maritime, meaning that the air masses formed over an ocean, and the "T" stands for tropical, meaning the air mass started over tropical latitudes. So the typical mT air mass started over the Caribbean and Gulf of Mexico. The other main types of air masses are "mP" and "cP," where the "c" stands for "continental" and the "P" stands for "polar." mP air masses start over the north Pacific and bring rainy, cool weather to the West Coast in winter. cP air masses bring the cool, dry winter weather to the Midwest. To find a cT air mass, you'd probably have to seek it out in the Sahara or central Australia, although small cT air masses sometimes sneak northward from Mexico.

30. The Weathermen, later known as the Weather Underground, were a radical offshoot of the Students for a Democratic Society in the late 1960s. Their name came from the Bob Dylan song "Subterranean Homesick Blues" which contained the line, "You don't need a weatherman to know which way the wind blows."

Plants

Some people think plants exist to produce pretty flowers and tasty fruits, while others—solemn geographers among them—believe plants are merely surrogate climatologic signatures, telling us what the weather is like in a region so we don't have to sit around measuring it.

The geography of plants varies with climate, evolutionary history, and human activity. Ever since humans learned plants are useful, we have fetched them from one part of the globe to another, bred them for human purposes until the original would be unrecognizable, and used them for purposes that would astonish anyone without a sense of humor. We grow plants, eat them, breed them, shape them, build with them, write on them, even alter the weather with them by cultivating some to shade the ground. We also use them as symbols—clear evidence of an object's importance in the human psyche. The most widely useful plant is probably the coconut palm. A display of palm products—from copra to soap to bows and arrows to houses—makes the point vividly at Fairchild Tropical Garden in Miami.

1. Where do willow trees grow?

2. Why do Venus flytraps eat insects? (Hint: "Because they're hungry" won't do.)

3. Where do kelp forests grow?

4. In 1992, 2,421 persons were admitted to U.S. emergency rooms for injuries involving h _____ s. Complete the partial word.

5. New England is thought by many to have the most glorious autumn colors in the world when the leaves change. People in England are so envious that they have begun importing an American plant to get the same effect at home. Can you name the English gardeners' plant of choice for fall colors?

6. The world's largest living thing is not the blue whale, but a plant, and it lives in the United States. What do you know about it?

7. The spotted owl came to be a symbol for a diverse ecosystem in the Pacific Northwest. What is the dominant tree in these old-growth forests?

8. What do "tallgrass," "shortgrass," and "mixed grass" refer to?

9. What region of the United States leads in commercial reforestation—that is, planting trees for eventual harvest?

10. What was the most valuable export of colonial America to Britain?

11. What has been described as "the only human activity with no downside"?

12. How many Americans are government-certified master gardeners?

    5,000 _____

    15,000 _____

    50,000 _____

    150,000 _____

    500,000 _____

13. Would it be better for your garden, when you die, if you were to be cremated and your ashes scattered over the soil, or buried there directly?

14. What "crop" occupies more square feet of land than any other in the United States?

15. Where in America can you find what Peter Farb called "whole bonsai forests" of trees that you can step over in a single stride?

16. How much water will a big tree lift out of the ground and into its topmost branches on a summer's day?

17. What does a pitcher plant eat, and how?

18. What "moss" is a member of the pineapple family?

19. Would you guess most of the plants that grow on high mountaintops are annuals or perennials? Why?

20. In the West, Englemann spruce trees and ponderosa pine trees have a similar look to the lay person's eye, but are very different in at least one respect—where they grow. Why can spruces grow in places where pines cannot?

21. Where in the United States will you find 250 varieties of flowering herbs that are found nowhere else?

22. If there weren't any oceans or mountains to influence plant life and how it grows, then sunlight, heat, and soil would be the only influences. In that case, how would we see the vegetation rearranging itself?

23. One of your coauthors once spent a night in a hammock 125 feet in the air in the top of a tulip poplar tree in the Joyce Kilmer State Forest in western North Carolina. Some of these trees grow to 200 feet in height, and are intermingled, at Great Smoky Mountains National Park, with twenty other species that reach record sizes, as well as 1,400 flowering herbs, 350 mosses, and 2,000 species of fungi. Can you mention some other quite remarkable things about this great regional forest?

24. You know that bittersweet feeling we may get when autumn arrives and the leaves begin to turn? What causes that? (The turning of the leaves, not our feelings.)

25. Can you arrange these plants in the order in which their leaves change in fall, with the earliest first?

Ash trees and sugar maples _____

Herbs and berry bushes _____

Red maples _____

Oaks _____

26. Early this century, the redwood was a great boon to the timber industry, entire churches being built from a single tree. John Muir, the naturalist, foresaw the threat to these magnificent giants, and he said: "No doubt these trees would make good lumber after passing through a sawmill, just as _____ _____ after passing through the hands of a French cook, would have made good food." Fill in the blanks.

27. How long is the leaf of a redwood tree?

28. Where is the only U.S. rain forest?

29. What is a "nurse tree"?

30. Why would corn die out completely if not for the help of humans?

31. How are those varicolored ears of corn you see in the fall created?

32. There are many theories to explain why the midwestern prairies remained prairies—have you ever thought about that? Why haven't they been invaded and taken over by trees, or shrubs, or something?

33. What did John C. Fremont, searching in 1844 for a mythical river that watered the desert, call "the most repulsive tree in the vegetable king-dom"?

34. What tree has been described as "a tree de-signed by someone who had never seen a tree" and "a tree turned inside out"?

35. What do these plants have in common: daisy, Queen Anne's lace, dandelion, mullein, chicory, black mustard, butter-and-eggs?

36. Where are the Pine Barrens?

37. Fourteen million trees. What does this number represent?

Plants

1. Willows, of the genus *Salix,* grow with their roots in or near groundwater nearly worldwide. Thus they are found along streams and at springs. Only a few species "weep."

2. Venus flytraps, plants of the southeastern United States, whose leaves snap shut when an insect alights, apparently get minerals from their bug dinners that the local soil does not provide.

3. Kelp is an oceanic plant that grows just outside the surf zone on much of the U.S. west coast. It provides a basis for one of the earth's most productive ecosystems.

4. Houseplants.

5. Poison ivy.

6. We used to think the largest living thing was "General Sherman," a giant sequoia tree 275 feet tall and weighing 2,750 tons. If split into matchsticks, the good general could supply a light to every person on earth. Then, for a long time we thought the world's largest living organism was a "creepy giant fungus," as someone once called it, that dwelt underground in northern Michigan. The thing had been there, growing silently, for somewhere between 1,500 and 10,000 years—nobody really knows how long—until it sprawled over 30 acres and weighed 100 tons. Pretty big. But now it is known—or anyway believed (we have a feeling this could be an endless shaggy-dog tale) that the creepy fungus is *not* the largest living thing after all. The LLT is actually a tree—or more accurately, a forest of 47,000 aspen trees growing from a single root system in the Wasatch Mountains of Utah. It covers 106 acres. It weighs 6,000

tons. And when the leaves change color in autumn, they all do it at exactly the same time.

7. The Douglas Fir.

8. These are the three types of prairie that originally spread across the Midwest and Great Plains of the United States. The tallgrass was in the more humid East, the shortgrass was in the drier West.

9. The Southeast leads in reforestation, particularly Mississippi, Alabama, and Georgia.

10. By a large margin, it was tobacco.

11. Gardening has been thus described. Bertrand Russell once said he concluded that happiness was possible for humankind only after a talk with his gardener.

12. Fifty thousand of us are certified master gardeners.

13. The garden would prefer to devour you raw.

14. Lawns. The total amount of turf grass exceeds wheat, corn, or tobacco. Homeowners use ten times more pesticide per square foot than farmers do and, as of mid-1993, thirty-two of the major thirty-four pesticides they employ had not been fully evaluated for their impact on humans.

15. On Cape Cod, where cold, salty wind prunes and dries the vegetation and robs the soil of nutrients. "The result," Farb says, "is a forest of pygmy trees, true-to-scale miniatures, produced in the same way as the bonsai trees grown by Japanese gardeners."

16. A big tree can lift a ton of water in a day, moving it at the speed of 150 feet per hour.

17. You might expect to see the pitcher plant in the "Little Shop of Horrors." One of our native carnivorous plants, it feeds on insects. Its leaves are curved into a pitcher shape and marked with handsome crimson veins and a red lip—like many flowers. An insect lands on the red lip, trustingly follows the crimson veins down into the cool, dim interior of the pitcher, at the bottom of which is a pool of water where, very often, float the remains of its predecessors. By then it is too late. The pitcher's neck is steeply angled and smooth as ice. The unfortunate bug slip-slides away, splashing daintily into the water, where enzymes break down its tissue so the plant can absorb the nutrients.

18. Spanish moss, so called, which dangles so atmospherically from the oak trees in *Gone with the Wind* and all over the humid South, is related to the pineapple. It is not a parasite, and takes its nutrients from the air, doing no harm to the live oaks it seems to prefer to dangle from. Though seemingly gray in color, Spanish moss is actually green beneath an outer skin, and it produces seed pods and yellow flowers so tiny that most Southerners have likely never seen one, or noticed it if they have.

19. Mountaintops are so cold, windy, thin-of-air and generally miserable that an annual plant wouldn't stand much chance of getting established. Only the hardiest perennials are adapted to these conditions.

20. Spruces have shallow roots, so they can survive in the infertile soil of high elevations. That's why you find them higher up than pines. Pines come next, and then, farther down, near the rivers, we find water-loving trees like alders, birches, and willows.

21. In the Alpine Zone of the Rocky Mountains. Such lovely alpine meadows can be seen in Rocky Mountain National Park.

22. Plants would start shifting around until they had arranged themselves in neat east-west bands across the continent, with no trees at all in the north, where it's too cold, then a band of tundra and stunted trees, then a great forest of conifers—pines, firs, spruces—and then deciduous forests of maples, oaks, elms, followed by some narrow strips, in the extreme south, of palms and jungly plants that thrive in warm weather and a lengthy growing season. This would be very interesting to see from the air.

23. The Smokies, because they are relatively inaccessible, have been spared much of the logging which has devastated America's other virgin forests in the Northeast and Pacific Northwest. And the Smokies, on their western slope, have the second highest rainfall of anyplace on the continent, right after the Pacific Northwest.

24. The leaves don't turn because of the frost. The turning comes, like our feelings, from inside the organism. When the days begin to shorten, the failing light tells the tree it is time to shut down for winter. The sap system must still itself; the leaves, which have been evaporating water all summer and carrying on photosynthesis, must be got rid of. There is a predestined spot on each leaf's stem where it will break off, and as winter approaches, the process begins. The tree stops producing chlorophyll, which slowly departs from the leaves so that we, in our cars or our rockers on wide porches, can now see the yellows and golds that were there all along. As fall progresses, the tree adds reds and lavenders, for reasons not completely understood. But the brighter and sunnier the autumn, the more brilliantly crimson will be the show of leaves. As winter sets in, the gush of hormones which have kept the leaves joined to their stems begins to taper off, and they begin to fall, one by one, faster and faster, as many as 10 million of them per acre.

25. The smaller plants turn first—the herbs and berry bushes—followed by the red maples, then the sugar maples and ashes, and finally the oaks.

26. George Washington.

27. A redwood tree leaf is about a quarter inch long.

28. In the Pacific Northwest: the Olympic rain forest. This is a temperate rain forest. There are remnants of tropical rain forests in Puerto Rico and the northeast side of the island of Hawaii.

29. When a tree falls in the Olympia National Park, its trunk, lying on the forest floor, becomes a host for new growth. The new seedlings, higher up off the ground, have a competitive advantage and thrive, producing a straight row of saplings growing from the decaying nurse tree's trunk.

30. Corn, especially the hybrid corn we cultivate for food (95 percent of the corn we grow is hybrid), has no natural method of seed dispersal. If an ear fell, each kernel might generate a seedling, but they would be huddled together in competition for food and water, and eventually would probably all die out.

31. By planting a lot of varieties of corn and letting them alone. If you plant different kinds of corn close together, they will cross-pollinate and produce variable, edible Halloween or Thanksgiving decorations. It's harder to produce the consistent types we see every day because the farmer has to keep the strain pure.

32. One theory is that prairie fires kept other vegetation out, by killing only a single year of grass growth at a time while leaving the living roots to reseed and resprout. Fire often kills a tree entirely. Another theory is that the matted roots of a long-time prairie are so thickly intertwined that they prevent any other plants from getting a start among them.

33. The Joshua tree, with its spiny fingers and wildly waving arms, looks like the vegetative equivalent of a big-city street loony. You want to cross to the other side of the desert and avoid it. It got its name from the Mormons crossing the desert, to whom its mad "gesticulations" brought to mind an Old Testament patriarch urging on the children of Israel.

34. The saguaro cactus, whose enormous fluted column supports spiny, angled arms. These odd-looking but now-familiar shapes are perfect for storing up to a ton of water after a rare desert downpour, their trunk-pleats expanding accordionlike to make room. After a heavy rain, 95 percent of a saguaro's weight may be water. The portion of a saguaro that handles water and sap is on the outside; inside is a woody skeleton.

35. All are "weeds"—and all originated in Europe. They were brought here and distributed in the pants cuffs of settlers or attached to the fur of their animals. Of course, the word "weed" is open to interpretation.

36. The Pine Barrens are in the southern part of the nation's most densely settled state, New Jersey. The region is barren because its soils are not fertile enough for farming, and have thus been avoided since colonial days. Pines, however, thrive in the barrens.

37. The number of trees it takes to produce the mail-order catalogs sent to U.S. households each year.

QUIZ 6

Animals

Unlike plants, which simply stand or sprawl there soaking up sunlight, water, carbon dioxide, and minerals, animals travel to get what they need. Thus, discovering their geography isn't always easy; they may not be where you saw them the last time, and even if you do spot a particular species again and again, it may not be the same individual you saw there before.

A good way to study the geography of animals is to link them to the places they seem to like, then try to discover why they prefer these places and how they modify the places. What do they eat and what eats them? How do they use the landscape to resist predators, or to hide from them? Do they move around during the day, or at night? Do they migrate, following the availability of resources they need? Do they live in groups, like deer, or skulk around as individuals, like cougars? Are they common across a wide area? How do animals change their territory? And how does all this change when humans enter the picture? How have animals changed our lives, how have we changed theirs? Who benefits?

1. Do our pets tell us something about ourselves?

2. For every pound of humans living in the United States, how many pounds of bugs live here?

    3 pounds _____

    30 pounds _____

    300 pounds _____

3. How many abandoned pets are killed each year in American animal "shelters"?

4. What animal helped determine the routes of early American railroads? How?

5. What was the most common American bird in the mid-1800s?

6. If your dog or cat seems to spend most of its time sleeping and eating, what's wrong with it?

7. Here is a list of the top dogs in popularity, with the number of each breed registered with the American Kennel Club in 1992. Can you match the breed with the number registered, thus ranking them in order of popularity? (Hint: The number-ten dog is the chow, with 42,670.)

120,879	German shepherd
95,445	Beagle
91,925	Labrador
76,941	Shetland sheepdog
73,449	Rottweiler
69,850	Dachshund
60,661	Golden retriever
50,046	Poodle
43,449	Cocker spaniel
42,670	Chow

8. In New England, it is a darning needle. In Mississippi, it is a mosquito hawk. What do most of the rest of us call it?

9. Can you pick the fastest-running U.S. animal?

    Cow _____

    Thoroughbred horse _____

    Pronghorn antelope _____

    Cheetah _____

    Puma _____

    Yellow runner _____

    Black skimmer _____

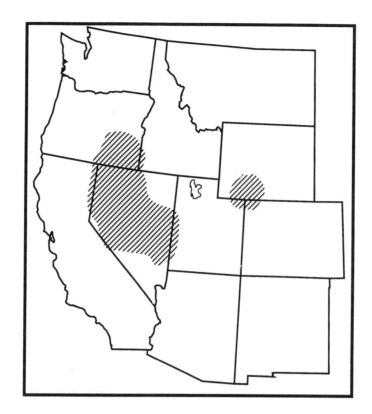

Here are two regions of wide plains and isolated mountain ridges in the Wild West. The country's major concentrations of what wild animal can be found in these small areas?

Answer: Wild horses. (Thanks to geographer Richard Symanski for suggesting the question.)

10. What was probably the last of the large mammals to find its way from Asia to North America across the land bridge of the Bering Strait?

11. For what part of its body does the green turtle get its name?

12. Most tropical animal species have slowly been driven southward by the human population into warmer, less populated zones, as humans thickly settled the Sunbelt. Can you name an exception?

13. It may surprise you to know that one "wild" animal is more numerous today than before Europeans arrived in North America, but it's true. Can you name the animal? (Hint: It isn't the wolf.)

14. What do these animals have in common: house sparrow, pigeon, Norway rat, starling, house mouse, brown trout, carp, pheasant?

15. Where did the common pigeon come from?

16. What is unusual about the Key deer and where does it live?

17. Which North American animal historically had the greatest range?

18. There is only one innately poisonous lizard in the world, *Heloderma suspectum,* and it lives in the southwestern U.S. desert. What is its common name?

19. What is the biggest freshwater fish in the United States?

20. The California condor is the largest bird in America, in terms of wingspan. But what is the *tallest* U.S. bird?

21. America is home to the largest land-dwelling carnivore in existence. Can you name this "horrible" beast?

22. Early colonists depended greatly on fishing and timbering to earn a living, but soon another trade overshadowed these and came to dominate commerce in young America. What was it?

23. What island was the center of the American whaling trade, and why?

24. An illegitimate child born in Haiti in 1785 and named Jean Rabin became one of America's most famous men under another name. What was the name and what was he famous for?

25. What does Kirtland's warbler warble?

26. Where does the rare Devil's Hole pupfish live?

27. Of which class of animals have humans thus far failed to exterminate a single species in the United States?

28. What was the first of the U.S. national parks?

29. Who declared the following about what creature? "He is a bird of bad moral character; he does not get his living honestly. . . . Besides, he is a rank coward . . ."

30. Five of every ten houses and seven of every ten apartments in the United States have what kind of animal living in them?

31. A shocking new grasshopper from South America hopped into Florida in 1992. How big is it?

32. Where might you find white spiders and white lizards?

33. What were "the greatest animal congregations that ever existed on earth"?

34. Where and when did the last passenger pigeon die? What famous naturalist died in the same year?

35. What proportion of Americans give Christmas presents to their pets?

Animals

1. Yes, according to two psychological studies of humans. In one, male dog owners scored high in aggression and dominance, while male cat owners scored high in autonomy. Dog-people scored higher in nurturing than cat-people. Turtle or tortoise owners were hardworking, reliable, and upwardly mobile, while snake owners liked to break the rules and be different. Horse owners were assertive and introspective, but low in warmth and nurturing. Bird-lovers were expressive and socially outgoing. Female bird-lovers scored high in dominance. These results are revealing because, according to the study's coauthor, Robert Kidd, we look for the same thing in pets as we do in friends: similarity and complementarity. "As a psychologist," said Aline Kidd, the study's other author, "I've spent my life telling people stereotypes are garbage, but these aren't stereotypes—they're reality."

2. There are 300 pounds of bugs for every pound of us.

3. Ten million animals are put to sleep each year, usually after a Death Row stay of five to seven days in "shelters." A few years ago, the figure was 15 million. Many people, out of false thrift or misplaced sentimentality, still do not spay and neuter their pets.

4. The bison. Huge herds trampled trails across the West, taking the routes of least resistance. Many of these routes were plotted by railroad surveyors and followed when track was laid. By the 1920s, only a few-score bison remained in all America. Today, many thousands live in protected areas, and you can order bison steak in some restaurants. Its taste is virtually indistinguishable from that of very lean beef—yet the meat is lower in fat, cholesterol, and calories than beef, pork, skinless chicken, and even some fish. No wonder there are 1,000 buffalo ranches in the United States and 130,000 buffalo. "Animals that people eat do not become extinct," says the director of the American Bison Association. Three out of four of today's bison are raised for slaughter.

5. The passenger pigeon was the most common American bird, and the most numerous bird ever to exist on earth—so common in 1848 that a protection act was intended to shield people netting the birds from molestation, rather than the birds themselves. In one nesting, in Petoskey, Wisconsin, in 1878, approximately 136,000,000 pigeons settled into an area of 750 square miles. Hunters arrived with nets, guns, and traps, and wiped out millions of them, selling their corpses for 25 cents a dozen. The slaughter was on. The last wild pigeon was killed in Pike County, Ohio, only twenty-two years later, leaving just a few remaining in captivity. They soon died too.

6. Nothing, probably. Nearly all mammals spend 90 percent of their time sleeping or eating, or waking up hungry, or falling asleep full. The proportion varies with the animal. A meat-eating cat or dog can bolt twelve hours' worth of food in seconds, leaving plenty of time to snooze, while a cow, since ruminants depend on less-nutritious grass, has to spend more time eating, and digesting, and eating, and digesting. . . . The one mammal which does not fit this pattern—the one that works crazy hours and stays up late—is the one that has a mortgage.

7. Ranked in order of popularity, the dogs are: Labrador, rottweiler, cocker spaniel, German shep-

herd, poodle, golden retriever, beagle, dachshund, Shetland sheepdog, and chow. The list shows a growing preference for large, potentially aggressive guard-dog breeds. The hot rottweiler, which has been climbing the ranks for several years, pushed the cocker spaniel from second to third place. Labs had pushed cockers out of first place the year before and held on to first place again. Doberman pinschers, which were a great favorite in the 1970s when they rose as high as number two, dropped to twentieth place. The next fashionable big dog may be the akita, which ranked thirty-third but was moving up fast, growling and snapping as it went.

8. A dragonfly.

9. The chestnut-and-white pronghorn antelope, not truly an antelope since it is unrelated to Old World animals carrying this name, is the fastest-running animal in North America.

10. Humans were very likely the last significant mammal to arrive in North America, hot on the trail of the bison, elk, deer, moose, caribou, and others.

11. Green turtles are called green for the color of their edible fat, which made green turtle soup so delicious, which is one reason they are now nearly extinct. The related hawksbill turtle, also endangered, was the source of tortoise shell, now mainly supplanted by plastic.

12. The armadillo, once confined largely to the southern U.S. border regions, is now found in Oklahoma, Florida, and Georgia. The cattle egret, once an inhabitant of Asia and Africa, appeared in South America early in this century and has also been spreading northward ever since. In the 1950s, it was spotted for the first time in Canada. The bird has succeeded largely because no native species already occupied the ecological niche it prefers: It feeds on insects flushed by grazing cattle.

13. There are more white-tailed deer, the most abundant large mammal in North America, than there were when the continent was wilderness. We have depleted their predators, notably wolves, which had formerly thinned the deer herds. Now,

supplanting those predators, we are left with the job of managing the deer population.

14. All are exotic species which accompanied Europeans to the New World and thrived here. Only two—brown trout and pheasant—would be likely to be kept if we now had a choice.

15. The pigeon descends from a wild Eurasian rock dove.

16. The Key deer is a miniature version of the white-tailed deer, about the size of a Labrador retriever, whose range is confined to a refuge on Big Pine Key in the Florida Keys. Fewer than 300 remain, and those are being crowded by development. The deer's fondness for cigarette butts has not helped, tempting it onto U.S. 1 where it is often struck and killed by traffic. As more Americans give up smoking, however . . .

17. The puma, cougar, or mountain lion (different names for the same big cat) once ranged into every state in the union and as far south as Argentina. It has been destroyed by humans wherever found, even though its documented attacks on humans number less than a half-dozen in centuries, and it is threatened or endangered everywhere it remains. A subspecies, the Florida panther, numbers only a few dozen.

18. The Gila monster. It has a close relative in Mexico, *H. borridum.* Some giant monitor lizards such as the Komodo dragon have poisonous saliva but, in them, the toxin is a mix of bacteria.

19. The lake sturgeon is largest. A primitive species reaching 8 feet and 300 pounds, it is thought to be extinct from all the U.S. river systems in which it thrived, and survives, if at all, only in the Great Lakes.

20. The whooping crane, gleaming white with black wingtips and a mask of crimson, stands over 5 feet tall. Whooping cranes were so common in the United States that Audubon once killed seven with a single shot (in the days when naturalists killed the creatures they studied). By 1926, it was estimated that fewer than a dozen pairs remained in all of North America. Breeding programs have recently begun to increase their numbers.

21. The Kodiak bear of Alaska, one of which has been measured at 9 feet 2 inches tall and 1,656 pounds, is the largest terrestrial carnivore. It is simply a well-fed, jumbo version of the grizzly bear, *Ursus horribilis.*

22. The fur trade, which took millions of pelts from beavers, martens, fishers, bears, wolves, otter, and many other species.

23. Nantucket, off Cape Cod, was the center of whaling, not only for its central location but because its soil was so poor the settlers had little choice but to get their living from the sea. The whalers first went after the bowhead and right whale, the latter so-named because it floated when killed, making its retrieval easy. Later, after the Revolution, the sperm whale became most important, its oil being sold on the world market and becoming, as Peter Matthiessen observed, "the bulwark of the new nation's economy." The sperm whale, he said, "may have rivaled the beaver as the most significant wild creature in the history of North America."

24. Jean Rabin became Jean Jacques Fougere, then Jean Jacques LaForest Audubon, then John James Audubon, perhaps America's most famous naturalist and painter of the classic work, *Birds of America.*

25. Kirtland's warbler, with its yellow breast and blue back, breeds only in a confined area of Michigan's Lower Peninsula in groves of young jack pine which, to suit the bird, must be under 18 feet tall. Thus, Kirtland's warbler appears to need forest fires to survive, since only fires create the burned-over ground on which young jack pines sprout. Its call sounds, to Peter Matthiessen, like: "Felicity-has-to-wee-wee."

26. The Devil's Hole pupfish, a frisky little fish, lives in a 15-by-40-foot remnant of an Ice Age lake called Devil's Hole—where else? This may be, as one naturalist observed, "the smallest range of any vertebrate in the world." It is part of the Death Valley National Monument.

27. Difficult as it is to believe, given the proven destructive power of humans on other species, we have not managed to kill off a single species of reptile in America, not even a snake or a toad. This is certainly not true on other continents, however, where deforestation and pesticides and settlement have done irreparable damage. The same could easily happen here, if it hasn't already without our knowing it.

28. Yellowstone National Park, established in 1872, was the first. Yosemite and Sequoia were established eight years later, Mount Ranier and Crater Lake just after the turn of the century.

29. Benjamin Franklin wrote these words about our national bird, the bald eagle, which lives mostly on carrion. As for the wild turkey, the bird he favored for the post of honor, there are only twice as many of them left as there are turkey hunters.

30. Half the houses and seven tenths of American apartments are infested with cockroaches, according to Dr. Vernard Lewis, a leading entomological researcher at the University of California also known as Doc Roach. He defines a "moderate" roach infestation as 15,000 bugs. He defines a serious infestation as 100,000 roaches or more. Roaches carry more than sixty diseases and can eat virtually anything. "They'll eat the fingernails of babies lying in their cribs, and lick the milk from the corners of their mouths," says Doc Roach. "They'll nibble at the high-protein mucous solutions in the corners of a baby's eyes." Roaches' blood is green. Just thought you'd like to know.

31. The new grasshopper is 5 inches long. It has been seen licking its lips and eyeing Buicks.

32. You would find such creatures among the pale gypsum dunes of White Sands National Monument, where their colorlessness helps them hide from their enemies.

33. The herds of millions of bison—60 million by one estimate. They once grazed across North America, not only on the Great Plains but also in the eastern forests, in subtropical Florida, coastal Georgia, and along the Chesapeake Bay. They were slaughtered by Europeans and, almost as greedily and wastefully, by Indians, who had simply lacked the numbers and technology to do the job efficiently. Native Americans, who usually killed only heifers and cows, frequently took bison just for their tongues, and like the settlers later, left carcasses rotting all across the plains. It was a gang of settlers and Cree Indians together who destroyed the entire herd of northern bison in North Dakota in 1883. It was the last of the great herds. Later that same year, the North Dakota legislature passed a bison-protection measure.

34. The last passenger pigeon, named Martha, died in the Cincinnati Zoo in 1914, the same year naturalist John Muir died.

35. About 68 percent of us give our pets Christmas presents, according to a Gallup poll in 1990. Twenty-four percent of us celebrate their birthdays—which is more than they do for us.

The Environment

Environmentalism may have its roots in religion; most of the great belief systems speak of the earth as God's gift to the animals, including us, and of our stewardship thereof. Yet it is a rational movement as well. We have let the environment slide as we concerned ourselves with short-term gains. Now, as populations rise and technology multiplies our power to do damage, it is harder to ignore the implications of shortsightedness.

Environmental interaction is a fundamental theme of geography, and all real environmentalists are geographers, because they have to be.

1. Where is the quietest place in America?

2. What is the difference between a swamp and a marsh?

3. How many tons of waste did we Americans produce in 1992?

4. What does Garrett Hardin's first law of ecology hold?

5. Where does most of the water flowing down the Colorado River go? (Yes, this is a trick question.)

6. The Interior Department has the job of designating endangered species in America. In 1966, the first year of such designations, the list contained 78 species. By 1992, there were 749 endangered plants and animals on the list, and more were endangered all the time. By 1997, about how many endangered species were there expected to be in America?

7. Match these national seashores with their states:

Cape Hatteras	Massachusetts
Padre Island	Virginia
Assateague	Florida
Canaveral	California
Cape Cod	Texas
Point Reyes	North Carolina

8. Match these national monuments with their states:

Effigy Mounds	Colorado
Organ Pipe Cactus	California
Salinas Pueblo Missions	Idaho
Natural Bridges	Montana
Devil's Tower	Wyoming
Custer Battlefield	Utah
Craters of the Moon	New Mexico
Pinnacles	Arizona
Great Sand Dunes	Iowa

9. Associate these national forests with their states:

Apalachicola	California
Tombigbee	Idaho
Davy Crockett	South Carolina
Talladega	Oregon
Francis Marion	Florida
Huron	Arizona
White Mountains	Indiana
Allegheny	Montana

Daniel Boone	New Mexico
Ozark	South Dakota
Mark Twain	Missouri
Black Hills	Arkansas
Santa Fe	Texas
Tonto	Kentucky
Angeles	Pennsylvania
Willamette	New Hampshire
Boise	Michigan
Helena	Alabama
Hoosier	Mississippi

10. Where in the United States will you find tundra?

11. Environmentalists are always hollering about the evils of packaging. What was so terrible about the long box for compact disks or the fast-food Styrofoam hamburger-keeper?

12. Which of these materials makes up the largest portion of packaging waste?

Plastic _____

Paper and cardboard _____

Glass containers _____

13. True or false: We discard enough iron and steel to supply all manufacturers as much as they use.

14. True or false: 70 percent of all metal is used just once.

15. True or false: 3 million cars are abandoned in the United States each year.

16. True or false: It takes 65 percent less energy to recycle a metal can than to make a new one.

17. True or false: We spend about $200 million a year to heat our hot tubs.

18. True or false: Washing dishes by hand uses more water than washing them in a dishwasher.

19. Do you know all the advantages of using a fluorescent light bulb instead of an incandescent bulb? Name a few.

20. Which of the following accounts for 50 percent of our average personal energy use?

Transportation _____

Home space heating _____

Hot water heating _____

Cooking, refrigeration _____

Cooling, air-conditioning, lighting _____

21. What is a wind farm?

22. What do praying mantises, ladybugs, the bacterium *Bacillus thuringiensis,* and the mold called mycorrhizae have in common?

23. What proportion of the population lives in rural America?

24. Using one of those gasoline-powered leaf blowers for an hour puts out as much pollution as driving a car how far?

25. Alaska has the most federal-government-owned land. What state has the second-most?

26. What is the Palouse and where is it?

27. Where is the Copper Basin and why should you care?

28. There are 780,000 lawyers in the United States, writing memos, filing briefs, wadding up scratch paper and tossing it dramatically into their trash cans. How much paper does each of these lawyers use in a year?

100 pounds _____

200 pounds _____

750 pounds _____

1,000 pounds _____

A ton _____

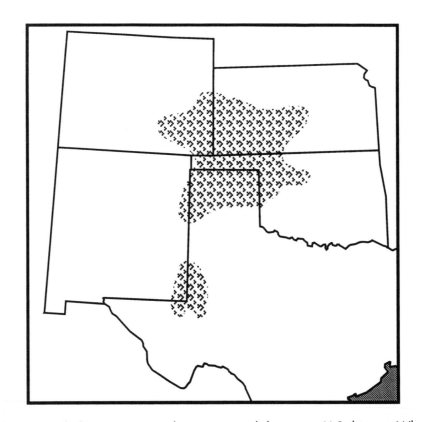

The pattern mapped here records the worst regional environmental disaster in U.S. history. What happened in these areas during the 1930s?

Answer: These were the legendary Dust Bowls of Depression America. They experienced severe drought during the early thirties—not surprising on the edges of deserts. But the regions has been plowed and farmed during the preceding decades, and probably should never have been brought under such extensive cultivation. In 1993, we learned similar lessons about building houses on the Mississippi floodplain. Construction along beaches, in hurricane and earthquake zones, is similarly risky.

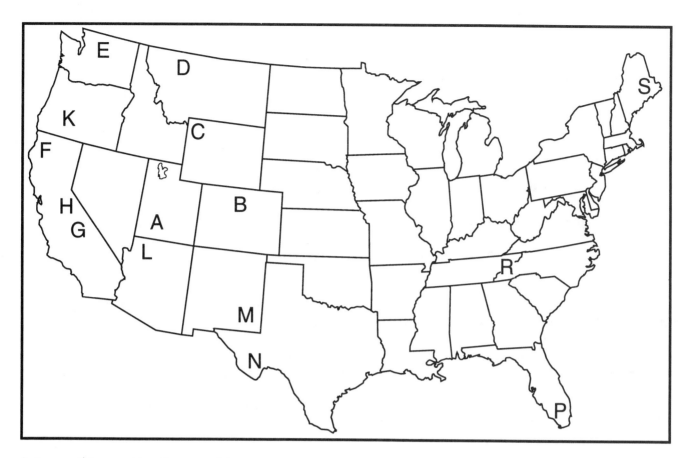

Letters on this map identify some of the country's most beautiful national parks. Using the list of parks below, locate each on the map.

Zion	Glacier	Big Bend
North Cascades	Sequoia	Grand Canyon
Carlsbad Caverns	Great Smoky Mountains	Rocky Mountain
Everglades	Acadia	Crater Lake
Yellowstone	Redwood	Yosemite

Answers:

P Everglades	A Zion	K Crater Lake
R Great Smoky Mountains	C Yellowstone	H Yosemite
S Acadia	B Rocky Mountain	G Sequoia
N Big Bend	D Glacier	F Redwood
L Grand Canyon	E North Cascades	M Carlsbad Caverns

29. The American Indian—great conservationist, right? Respecter of nature? Preserver of the natural legacy?

30. Who said, "Our lands were originally very good; but use and abuse have made them quite otherwise."?

31. In 1909, a report to President Theodore Roosevelt, an early conservationist, informed him that _____ _____ s of the nation's original timber had already been destroyed. Can you fill in the proportional amount?

32. What do the following items have in common? Unexploded bullets and bombs; radioactive runoff from a uranium mine; contaminated irrigation drainage; abandoned oil and gas wells; abandoned mines; construction dump sites contaminated with hazardous waste; leaking underground storage tanks; toxic runoff.

33. What animal introduced to North America by Europeans caused the most environmental change?

34. If horses lived in North America during the Ice Age—and they did—and then went extinct for reasons that are not known—and they did—how did they return?

35. Which of the lower forty-eight states has the most land covered with national forests?

36. Kudzu is famous all over the South. What is kudzu?

37. How long could you run a TV set on the energy we save by recycling a single aluminum can?

38. What is the worst threat to wild plant and animal species?

39. What was John Chapman's favorite apple?

40. What was the first year that Americans consumed more of the world's produce and goods, on average, than citizens of other industrialized countries?

1900 _____

1925 _____

1935 _____

1945 _____

1950 _____

1960 _____

41. What product, once restricted to clear mountain streams in the American West, spread via export to become one of America's notable contributions to world gastronomy?

42. Grant Wood, one of our country's greatest painters, is famous mostly for his oft-parodied "American Gothic." Yet he completed many landscapes of his native state that were almost surrealistic. What was his home state?

43. Arguably the greatest painter of seascapes who ever lived was an American who, like Grant Wood, had two first names (or maybe it was two last names). Can you name him and say where he painted his greatest oils of a rocky seacoast?

44. Nathaniel B. Palmer, an American sealer, was the first to reach a particular place in 1820 and thus give the United States the strongest claim to a vast territory which, however, it has not actually claimed. What territory are we speaking of?

45. Where is the world's tallest tree?

46. How much of the United States is forested today?

47. Here is a quotation from Aldo Leopold, one of the country's most important naturalists and conservationists: "Everybody knows that the autumn landscape in the north woods is the land, plus a red maple, plus a ruffed grouse. In terms of conventional physics, the grouse represents only a millionth of either the mass or the energy of an acre. Yet subtract the grouse and the whole thing is dead." Compose a little essay, on paper or in your head, or in conversation with a friend, on the subject of some particular living thing being subtracted from the landscape.

Answers

The Environment

1. The IBM Corporation's laboratory in Boca Raton, Florida, is a leading contender for quietest spot. For research into electronic materials and their performance, an echoless test chamber is required. This one eliminates 99.99 percent of all noise with walls covered with sound-absorptive fiberglass. Even the building's foundation is isolated from the surrounding property.

2. Swamps have mostly trees and marshes have mostly grasses. Both are flooded most of the time.

3. We produced 13 billion tons of waste in 1992—a lot to burn, recycle, or hide.

4. We can never do only one thing, is Hardin's law. Whatever we do has many other effects than the ones we intend, and many of these are unpredictable.

5. Most of the Colorado's water goes into the atmosphere, disappearing as water vapor evaporated into the air.

6. There were expected to be 1,150 endangered species in America by 1977.

7. Cape Hatteras, North Carolina; Padre Island, Texas; Assateague, Virginia; Canaveral, Florida; Cape Cod, Massachusetts; Point Reyes, California.

8. Effigy Mounds, Iowa; Organ Pipe Cactus, Arizona; Salinas Pueblo Missions, New Mexico; Natural Bridges, Utah; Devil's Tower, Wyoming; Custer Battlefield, Montana; Craters of the Moon, Idaho; Pinnacles, California; Great Sand Dunes, Colorado.

9. Apalachicola, Florida; Tombigbee, Mississippi; Talladega, Alabama; Francis Marion, South Carolina; Huron, Michigan; White Mountains, New Hampshire; Allegheny, Pennsylvania; Daniel Boone, Kentucky; Davy Crockett, Texas; Ozark, Arkansas; Mark Twain, Missouri; Black Hills, South Dakota; Santa Fe, New Mexico; Tonto, Arizona; Angeles, California, Willamette, Oregon; Boise, Idaho; Helena, Montana; Hoosier, Indiana.

10. You'd find tundra in northern Alaska. Tundra is the vegetation type found in the high arctic where the climate is too severe to support trees well. The main plants are mosses, grasses and a few stunted trees. Caribou live happily on the tundra.

11. Simply in cost terms, packaging accounts for 13 percent of food prices and 50 percent of the cost of dealing with garbage. In 1993, disposable packaging and containers made up nearly 32 percent of all American solid waste—57 million tons of it a year.

12. Paper and cardboard comprise 51 percent of packaging waste, glass containers 23 percent, and plastics only 13 percent.

13. True.

14. True.

15. True.

16. True.

17. True.

18. True, by about 6 extra gallons per load.

19. Not only does one 18-watt fluorescent bulb produce as much light as a 75-watt incandescent

bulb, but it lasts ten times as long, keeps 250 pounds of carbon dioxide out of the atmosphere, and leaves more than 80 pounds of coal in the ground. One bulb!

20. Transportation uses half of our average personal energy expenditure.

21. A wind farm is an array of windmills or, more properly, wind turbines, set up in a windy place to produce energy.

22. They are biological controls, natural agents used to control agricultural pests.

23. About one fourth of Americans live in the rural parts of the country, or outside of the metropolitan regions.

24. The use of a leaf blower pollutes as much as a car being driven for 100 miles, according to the Environmental Protection Agency. Lawn mower use equals 50 miles, and chain saw use 200 miles.

25. Nevada has the second-most government-owned land—79 percent of the state. In Rhode Island, the feds own only 5 acres, less than 1 percent of the state.

26. The Palouse is a region of rich soil derived from windblown silt in southeastern Washington State. It is famous for its highly successful wheat production.

27. The Copper Basin, in the southeastern corner of Tennessee, would make a good monument on how not to manage the environment. It is the center of 56 square miles of bare, severely eroded land, which supports virtually no plant life, animals, or even soils. Copper Basin was America's first great area of copper mining in the nineteenth century. Vast pits were excavated and open-air smelters, the technology of the time, were built. The result was a dead zone. Restoration programs underway since the 1930s have made progress, but the Copper Basin remains a useful object lesson.

28. Each American lawyer goes through a ton of mostly virgin paper every year, according to the

American Bar Association Journal. But while lawyers routinely recycle their arguments, one observer pointed out, they seldom recycle their paper. If attorneys were required to use recycled paper, it would spare 1.9 million trees a year, according to Sierra Club figures.

29. No more than the rest of us. Native Americans burned vast forests to the ground just to flush the game. They depleted the soil with careless agriculture and then moved on to level more forest. They killed thousands of bison unnecessarily, taking only the tongue and leaving the rest of the carcass to rot. When the white folks arrived, the continent looked pristine only in contrast to an overcrowded Europe. The Indians, many of them addicted to intertribal warfare, had kept their own populations so low their depredations were limited. They were no more self-absorbed and mindless than other humans, but no more sensitive, either.

30. George Washington, suggesting how early the trouble began.

31. Two thirds of the country's timber had already been destroyed by 1909, according to that report.

32. These pollution hazards are all inside America's national parks, Indian reservations, and wildlife refuges.

33. The hog, without doubt. The pig, *Sus scrofa,* can have three litters of up to thirteen piglets a year. They eat nearly everything. Huge numbers were imported on nearly every colonial ship. They readily escaped farms and ranches to establish feral populations and chewed their way across the New World.

34. Ponce de Leon apparently brought the horse back, from Puerto Rico to Florida, in 1521.

35. California has the most in the lower forty-eight. Alaska has 2 million acres more.

36. Kudzu is a vine native to Africa. It was brought to the United States in 1911 as a fast-growing livestock forage that would also combat soil erosion. It has now spread across the entire South, engulfing abandoned houses and fields. In the rural South, kudzu is the weed of weeds. (Recent tentative research suggests a kudzu ingredient may be useful in treating alcoholism.)

37. You could run a TV set about three hours on the energy we save by recycling a single beer can.

38. Destruction and alteration of habitat.

39. Johnny Appleseed, as he is also known, traveled western Pennsylvania, Ohio, and Indiana for forty years planting and tending apple seeds. This was not entirely a public-spirited endeavor, as he sold the saplings. His favorite apple was the Rambo. It is just barely possible that a tree he planted may still survive somewhere in the Midwest.

40. Americans were already the world's leading consumers of energy and material goods by 1900.

41. The rainbow trout is now commercially grown so successfully in so many foreign lands that much of the trout we eat in the United States is im-

ported. But trout is never as good as it is when cooked beside the stream it swam in.

42. Grant Wood was from Iowa, birthplace of one of your coauthors.

43. Winslow Homer did woodcuts in New York City and watercolors in the Caribbean and Florida Keys, but his greatest work—the intense, brooding oils—was done on the coast of Maine.

44. Antarctica.

45. The world's tallest tree is a redwood in the 58,000-acre Redwood National Park, established in 1968 along the Pacific coast of California. It is 369 feet tall.

46. Perhaps surprisingly, about 30 percent of the United States is forested today.

47. This is the only essay question in the book. If you gave it due thought, your answer is certainly correct.

Regional Map Quizzes

Geographers use the idea of regions as a tool to better understand a given part of the landscape. Some regions, such as "New England" and "The South," are primarily historical, evoking memories of Revolutionary War monuments or images of pickup trucks flying the Confederate flag. "The Great Plains" is more an environmentally defined region—a smooth, flat landscape covered with amber waves of grain. "The Midwest" is sometimes seen as a cultural metaphor for a nice, unpretentious, American common-sense ordinariness.

Here is a map quiz of states and important American cities presented by region—seven of them, divided simply for cartographic convenience. The idea is to locate the places by matching letters with states and numbers with cities.

Note that on the West Coast map, you are also asked to locate Cascade peaks marked with double-letters. (If you succeed at that, give yourself extra credit.)

On the Alaska and Hawaii maps, name the cities marked by numbers.

In Alaska locate the major national parks from this list: Wrangell-Mount St. Elias, Katmai, Lake Clark, Glacier Bay, Gates of the Arctic, and Denali. They are labeled with capital letters.

And in Hawaii, identify the major islands, labeled with capital letters, from this list: Molokai, Oahu, Niihau, Hawaii, Maui, Lanai, Kaui.

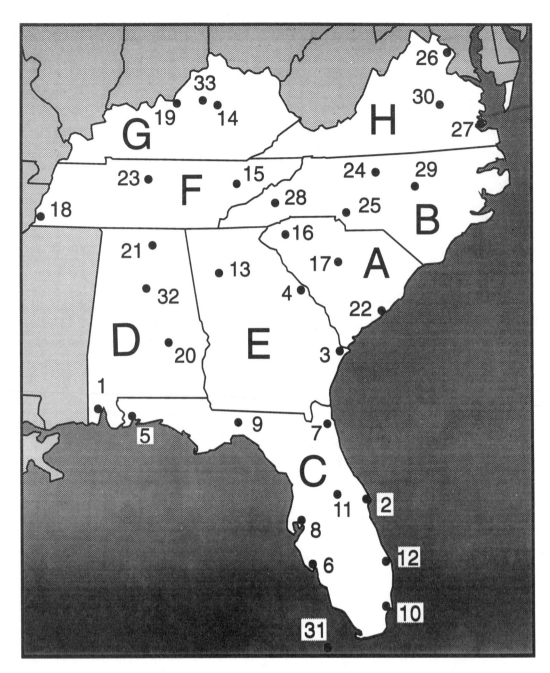

Answers:

SOUTHEASTERN STATES

C Florida	E Georgia
D Alabama	A South Carolina
F Tennessee	G Kentucky
B North Carolina	H Virginia

SOUTHEASTERN CITIES

5 Pensacola	9 Tallahassee
7 Jacksonville	11 Orlando
8 Tampa	6 Fort Myers
10 Miami	12 Palm Beach

4 Augusta	13 Atlanta
3 Savannah	32 Birmingham
21 Huntsville	20 Montgomery
1 Mobile	19 Louisville
14 Lexington	18 Memphis
23 Nashville	15 Knoxville
2 Cape Canaveral	17 Columbia
22 Charleston	16 Greenville
25 Charlotte	29 Raleigh
28 Ashville	24 Greensboro
30 Richmond	27 Newport News
26 Arlington	31 Key West
33 Frankfort	

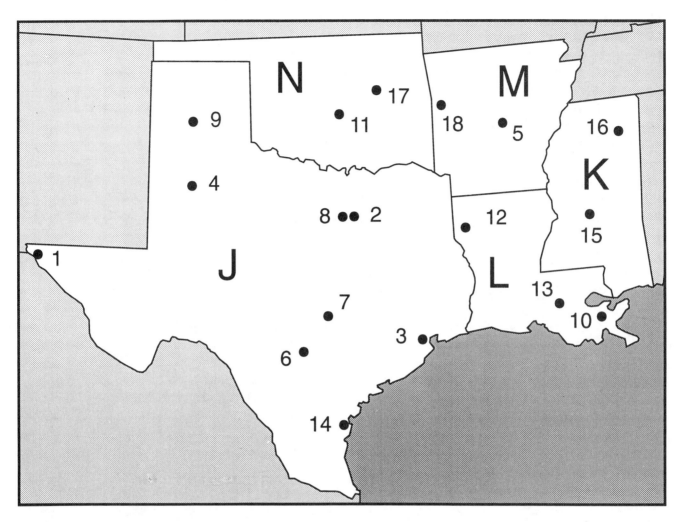

SOUTH CENTRAL STATES

J Texas   N Oklahoma

M Arkansas   L Louisiana

K Mississippi

SOUTH CENTRAL CITIES

10 New Orleans   13 Baton Rouge

12 Shreveport   11 Oklahoma City

17 Tulsa   5 Little Rock

6 San Antonio   3 Houston

7 Austin   2 Dallas

8 Fort Worth   1 El Paso

9 Amarillo   4 Lubbock

14 Corpus Christi 15 Jackson

16 Tupelo   18 Fort Smith

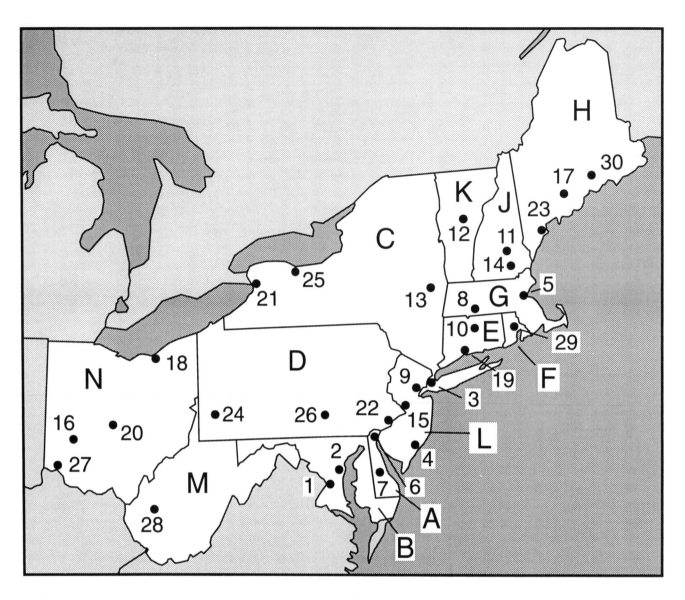

NORTHEASTERN STATES

N Ohio   M West Virginia

D Pennsylvania   C New York

B Maryland   A Delaware

L New Jersey   E Connecticut

F Rhode Island   G Massachusetts

K Vermont   J New Hampshire

H Maine

NORTHEASTERN CITIES

1 Washington, D.C.   7 Dover

6 Wilmington   2 Baltimore

5 Boston   8 Springfield

10 Hartford   29 Providence

14 Manchester   11 Concord

12 Montpelier   9 Newark

15 Trenton   4 Atlantic City

3 New York   13 Albany

21 Buffalo   25 Rochester

24 Pittsburgh   22 Philadelphia

18 Cleveland   20 Columbus

27 Cincinnati   16 Dayton

26 Harrisburg   19 New Haven

23 Portland   17 Augusta

28 Charleston   30 Bangor

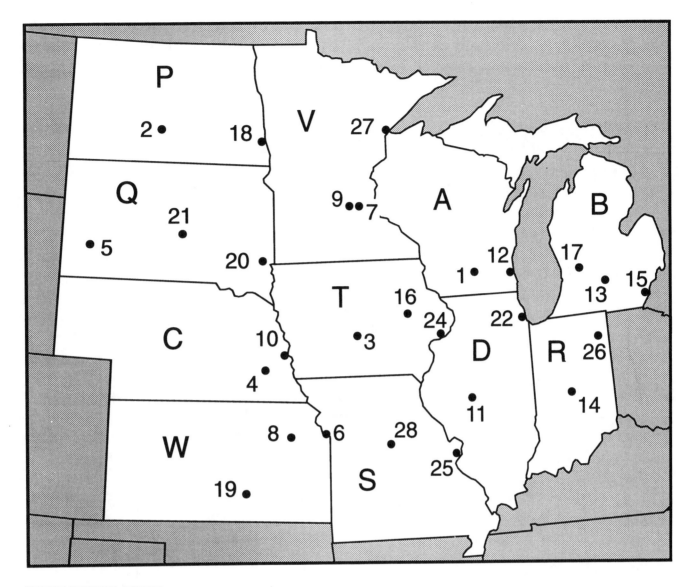

NORTH CENTRAL STATES

P North Dakota Q South Dakota

C Nebraska W Kansas

V Minnesota T Iowa

S Missouri A Wisconsin

B Michigan D Illinois

R Indiana

NORTH CENTRAL CITIES

2 Bismarck 18 Fargo

21 Pierre 5 Rapid City

20 Sioux Falls 10 Omaha

4 Lincoln 8 Topeka

19 Wichita 9 Minneapolis

7 St. Paul 27 Duluth

1 Madison 12 Milwaukee

22 Chicago 11 Springfield

23 Peoria 3 Des Moines

16 Cedar Rapids 24 Davenport

6 Kansas City 25 St. Louis

14 Indianapolis 26 Fort Wayne

15 Detroit 13 Lansing

17 Grand Rapids 28 Columbia

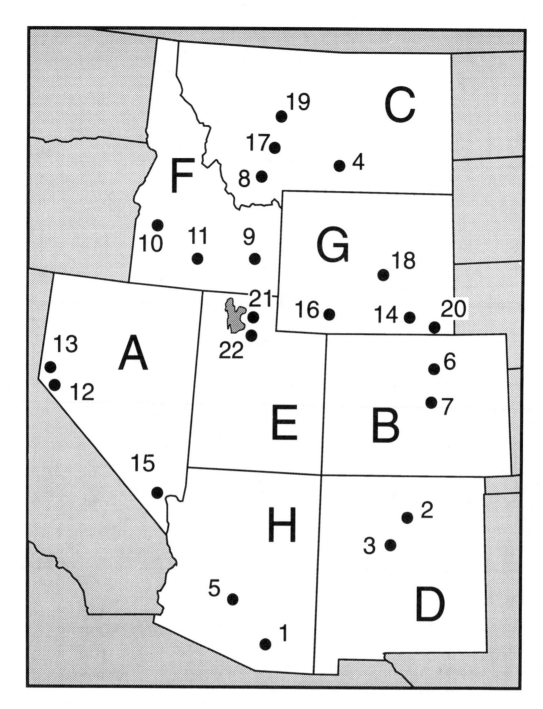

INTERMOUNTAIN STATES

C Montana   A Nevada

F Idaho   E Utah

B Colorado   H Arizona

G Wyoming   D New Mexico

INTERMOUNTAIN CITIES

4 Billings   13 Reno

8 Butte   15 Las Vegas

19 Great Falls   12 Carson City

17 Helena   5 Phoenix

20 Cheyenne   1 Tucson

16 Rock Springs   3 Albuquerque

18 Casper   2 Santa Fe

14 Laramie   6 Denver

9 Pocatello   7 Colorado Springs

11 Twin Falls   22 Salt Lake City

10 Boise   21 Ogden

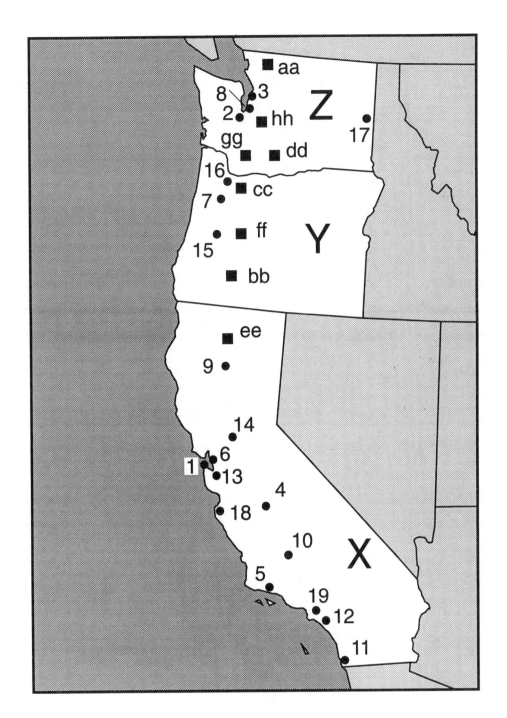

WEST COAST STATES

X California   Y Oregon

Z Washington

WEST COAST CITIES

3 Seattle   10 Bakersfield

8 Tacoma   1 San Francisco

17 Spokane   13 San Jose

2 Olympia   6 Oakland

16 Portland   11 San Diego

15 Eugene   12 Anaheim

7 Salem   5 Santa Barbara

9 Redding   18 Monterey

14 Sacramento 19 Los Angeles

4 Fresno

CASCADE PEAKS

ee Mount Shasta   bb Crater Lake (right; not
                        a peak)

cc Mount Hood   ff Mount Bachelor

gg Mount St. Helens hh Mount Ranier

dd Mount Adams   aa Mount Baker

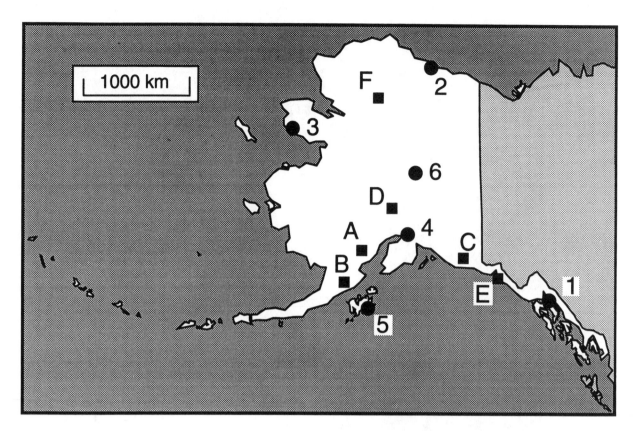

ALASKA CITIES

1 Juneau   3 Nome

4 Anchorage   2 Prudhoe Bay

6 Fairbanks   5 Kodiak

ALASKA NATIONAL PARKS

D Denali   B Katmai

A Lake Clark   E Glacier Bay

F Gates of the Arctic   C Wrangell-Mount St. Elias

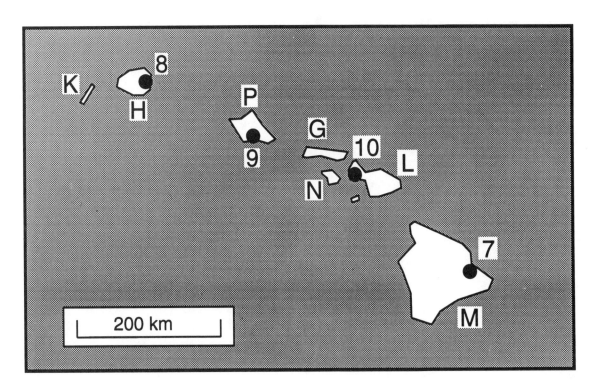

HAWAII CITIES

7 Hilo   9 Honolulu

10 Lahaina   8 Kapaa

HAWAIIAN ISLANDS

M Hawaii   H Kaui

L Maui   K Niihau

N Lanai   P Oahu

G Molokai

II

AMERICA'S HUMAN LANDSCAPE

The Edge: Borders

Borders divide things, separating regions and places from one another, insides from outsides. There are borders between states, between countries, between natural systems. Ocean and continent are both split and united by a seashore; two species of squirrel that have evolved from a common ancestor live on the north and south sides of the Grand Canyon. The latter is a clear example of a border across which something—in this case, squirrel genes—could not flow. The isolated creatures of the Galápagos Islands are a more familiar example.

Even without these visible, measurable kinds of borders, humans and animals often stay within territories of their own, marked with squirts of urine, gang graffiti, or international boundaries, defending them vigorously with threats, barks, cries, and wars. One of the most conspicuous borders in the world is marked by the Great Wall of China, perhaps the grandest public works project ever.

While borders divide, they can also unite in an odd way. Good fences make good neighbors, Robert Frost said.

1. Where was the following said, and by whom: "All border towns bring out the worst in the country"?

2. What does the Mason-Dixon Line divide?

3. What portion of Americans live within an hour's drive of the seashore?

    Under a fourth _____

    Over a fourth _____

    Over half _____

    Over three fourths _____

4. What do the northern borders of Wyoming, New York, and Vermont have in common?

5. What proportion of the Atlantic shore is "publicly owned and available for recreational use"?

    About 6 percent _____

    About 36 percent _____

    About 46 percent _____

    About 56 percent _____

    About 86 percent _____

6. What is, or are, "metes and bounds"?

7. What was the land division method of the Mexican colonization of California?

8. Why is so much of the United States a checkerboard of towns and landholdings laid out in a grid?

9. What is a township?

10. What four states join at the Four Corners?

11. Where is the Erie Triangle?

12. What is the "exclusive economic zone"?

13. What is the nearest foreign country to the United States that does not share a land border with the United States?

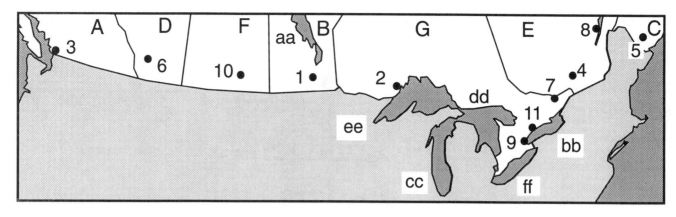

This map shows the southern part of Canada, our northern neighbor. Can you identify the Canadian provinces, major cities, and lakes along the northern border of the United States? Provinces are capital letters, cities are numbers, and lakes are double letters. You may choose from these lists if you need them.

Ontario	Vancouver	Toronto
Quebec	Fredericton	Hamilton
British Columbia	Calgary	Lake Winnipeg
New Brunswick	Ottawa	Lake Ontario
Alberta	Quebec	Lake Erie
Saskatchewan	Regina	Lake Huron
Manitoba	Winnipeg	Lake Superior
Montreal	Thunder Bay	Lake Michigan

Answers:

A British Columbia	6 Calgary	8 Quebec
D Alberta	10 Regina	5 Fredericton
F Saskatchewan	1 Winnipeg	bb Lake Ontario
B Manitoba	2 Thunder Bay	ff Lake Erie
G Ontario	9 Hamilton	dd Lake Huron
E Quebec	11 Toronto	cc Lake Michigan
C New Brunswick	7 Ottawa	ee Lake Superior
3 Vancouver	4 Montreal	aa Lake Winnipeg

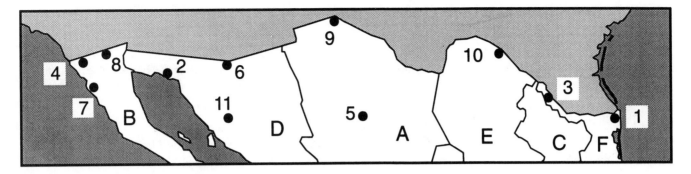

This is a map of the northern part of Mexico, our southwestern neighbor. Identify the states and major Mexican cities along this U.S. boundary. States are capital letters and cities are numbers. You may refer to these lists to jog your memory.

Nuevo Leon	Coahuila	Ciudad Juárez
Tamaulipas	Mexicali	Nogales
Baja California Norte	Ensenada	Puerto Penasco
Sonora	Tijuana	Nuevo Laredo
Chihuahua	Matamoros	Piedras Negras

Answers:

B Baja California Norte	F Tamaulipas	6 Nogales
D Sonora	1 Matamoros	2 Puerto Penasco
A Chihuahua	5 Nuevo Laredo	8 Mexicali
E Coahuila	10 Piedras Negras	4 Tijuana
C Nuevo Leon	9 Ciudad Juárez	7 Ensenada

14. What does 49 degrees north latitude represent in North America?

15. The United States has about 12,000 miles of "general coastline." More than half of this is in a single state. Which state?

16. How often do Siamese twins—more sensitively or politically correctly known as conjoined twins—occur in America?

17. Did the American Indians know where they were?

18. What, according to the U.S. Census Bureau, are the "outlying areas"?

19. When was the first map bearing the name America published?

20. The world's largest human construction is the Great Wall of China. The second-largest is in the United States. Can you name, locate, or describe it?

21. See if you can fill in these blanks. Alaska is the _____ est, _____ ernmost, _____ ernmost, and _____ ernmost state.

22. By 1775, how far inland from the coast had settlement of the original thirteen American colonies extended?

23. Eighty percent of _____ live within 10 miles of the U.S. border. Fill in the blank.

24. What do the northern borders of Arizona, New Mexico, and Oklahoma have in common?

25. What do Iowa, Missouri, and Arkansas have in common?

26. What do Minnesota, Michigan, Wisconsin, Illinois, Indiana, Ohio, Pennsylvania, and New York have in common.

27. What five states have panhandles?

The Edge: Borders

1. The Charlton Heston character says this in Orson Welles's 1958 classic movie, *A Touch of Evil,* which takes place in a Mexico-United States border town that looks a lot like Tijuana.

2. The original Mason-Dixon Line divided 230 miles of the border between Pennsylvania and Maryland. This boundary, at 39 degrees 43 minutes north latitude, was surveyed in 1767 by Charles Mason and Jeremiah Dixon. It is now a good example of what geographers call a relic boundary, a former political boundary which has lost its relevance. The people on both sides of the border are much more alike than different.

3. More than half of Americans live within an hour's drive of the shore.

4. They are all at 45 degrees north latitude, halfway between the Equator and the North Pole.

5. Only 6 percent of the beaches, marshes, and wetlands of the Atlantic coast are available for public use.

6. The term refers to the original colonial system of land survey, which was based on the simple expedient of striking lines between landmarks such as rocks and stream junctions. Those who live in the lands of an original colony, or in parts of Tennessee and Kentucky, or parts of Spanish Texas and California, can still see this system of irregular land division in the surrounding landscape. In the long run, however, metes and bounds have not worked out very well, since rocks can be moved, trees can die, and streams can be diverted.

7. Land grants were awarded to prominent settlers—the well-to-do and politically astute—during the 1820s to 1840s. These grants were enormous, and based on "improvements" (roads, buildings, corrals, etc.) described on artistic and sometimes inventive maps called *disenos*. Many of the grant holders lost a good bit of their land to lawyers when the titles were brought into the system of the United States after California statehood.

8. The Founding Fathers wanted to prove the power of their modern, scientific government over the land. So the system of rectangular land survey was used rather than the metes and bounds system. The initial legislation was the Land Ordinance of 1785, so that lands settled after this date follow the plan, which reached its apex of absurdity when a rectangular grid of streets was superimposed on the hills of San Francisco.

9. The township is the basic block of the United States land survey system, a square six miles on each side, or 36 square miles.

10. This oddity occurs only where Colorado, Utah, New Mexico, and Arizona connect.

11. Before the 1780s, Pennsylvania had no shoreline on Lake Erie. The state negotiated this narrow, 324-square-mile triangle from New York and Massachusetts and finalized the deal with cash in 1792. The city of Erie, crown jewel and centerpiece of the triangle and birthplace of one of your coauthors, is Pennsylvania's third-largest city.

12. This is the 200-mile-wide area along the American coasts that is claimed for exclusive use by the United States. In area, it is nearly as large as the land.

13. Russia. In the Bering Strait, less than 3 miles of water separate the islands of Little Diomede, Alaska, and Big Diomede, Siberia.

14. Most of the east-west border between the United States and Canada follows this line of latitude. It is said to be the longest undefended international border in the world. In much of the wooded West, it is marked by a cleared corridor 20 feet wide.

15. Alaska, with more than 6,500 miles of general coastline. "General coastline" is what you'd expect: the coast viewed in a general way, rather than measured in its every tiny twist and turn.

16. One in 60,000 pregnancies is with conjoined twins. The most crucial "borders" of these babies—those which separate individuals and create individuality—have been accidentally merged by nature. Something to think about when someone says, "It's natural, so it's got to be good for you." Only 30 percent are born alive or survive beyond one day.

17. Some did. According to the Spanish explorer, Francisco Coronado, Pawnees were using stellar maps marked on deerskins.

18. This is the formal name given to U.S. jurisdictions that don't fall under more conventional categorizations such as "state" or "territory." There are five: Puerto Rico, American Samoa, Guam, the Virgin Islands, and the Northern Mariannas Islands.

19. In 1507, by a German cartographer named Martin Waldseemuller. It was not very accurate, being based largely on hearsay and other secondary reports. The country's name comes from the Italian navigator, Amerigo Vespucci, who claimed that he reached the east coast of what is now named for him before Columbus got lost and found in the Caribbean.

20. The world's second-largest human construction is a dump—the Fresh Kills Landfill on Staten Island, in New York City. When "finished" in 2005 it will occupy 7.5 square miles and stand 505 feet above sea level, making it the highest terrain feature on the eastern seaboard. (Many buildings are taller, but not as massive.) If it were allowed to grow any higher, it would interfere with planes landing at Newark Airport.

21. Alaska is the coldest, northernmost, easternmost, and westernmost state. Odd but true. Isn't geography fun?

22. On the average, settlement had progressed only 70 miles inland, and the roads were terrible. The colonies still had closer contact with England than with each other. Native Americans, of course, had trails farther inland for centuries.

23. Canadians.

24. They are all 37 degrees north latitude.

25. They all have the Mississippi River as their eastern border.

26. They all have a border on one of the Great Lakes.

27. Texas, Florida, Oklahoma, Idaho, and Alaska have panhandles.

Place Names

How we name places is one of the most telling elements of geography. There are more than 4,000,000 geographic names in use in the United States. We have many ways of arriving at them. We may commemorate someone, as in Johnson City, or Washington, D.C. Or commemorate a place we remember fondly (even though we left it), as in New Amsterdam or New Hampshire. Or express a hope, as in Happyvale. Or codify a description, as in Fairview (or Buena Vista). Or stick with an Indian name, as in Lake Okeechobee or Yuma. Or let some taste-impaired real-estate developer name it something pretentious or irrelevant. (In Miami, there was a yeastily fragrant bakery on the corner of Red Road and U.S. 1—an olfactory landmark of the city for decades. A developer tore it down and erected a truly hideous shopping center on the site. He called it the Bakery Center. Not surprisingly, many Miamians celebrated when the Bakery Center went into receivership. Still, the scent is only a memory.)

Place names follow from our need to subdivide the world into manageable, apprehensible components. To geographers, they are more than labels, they are a matter of culture and history. They can also be useful, telling us something about where places are, how to get there, and what they meant to the people who arrived first.

1. Why are blue point oysters known as blue point oysters?

2. In 1784, Thomas Jefferson headed a committee charged with making policy for the then "western lands." What did he suggest as the name for what became Ohio?

3. What does "toponymy" mean?

4. In what state did a town change its name to Joe, hoping to lure football-loving tourists?

5. What New England state has an English name?

6. What do Rome, Syracuse, Ithaca, and Troy have in common, besides being in New York?

7. What was San Francisco's original name?

8. The source of the Mississippi River is a place called Lake Itasca in Minnesota. From which language does Lake Itasca get its name?

9. In what state are these places:

Pekin _____

Memphis _____

Babylon _____

Versailles _____

Cairo _____

Paris _____

Palestine _____

Moscow _____

10. Jack Dempsey, perhaps the greatest of America's champion boxers (he held the world heavyweight title from 1919 to 1926) was billed as the "Manassas Mauler." Where is Dempsey's hometown?

11. Where are these body parts to be found in the United States?

Wounded Knee _____

Finger Lakes _____

Cheek _____

Braintree _____

Finger _____

Hand _____

Heartstrong _____

Headville _____

Throg's Neck _____

Mouth of Wilson _____

Boca Raton _____

Boca Grande _____

Boca Chica _____

Muscle Shoals _____

Organ _____

Butte _____

12. Where are Hope, Charity, and Cape Fear?

13. Where are Fayettevilles?

14. Here are some place names in their original Indian languages, Translate them into current English and you'll know where they are:

"Upon a great plain"

"Place of skunk cabbage"

"Very big river"

"The place of great drunkenness" or "The place where we all got drunk"

"Canoe river"

"The red man"

"Upstream people"

15. Match the place with its Spanish word or phrase of origin:

Amarillo	Reindeer
Arizona	Strolling place
Colorado	Dovecote
El Paso	High pole

Florida	Snowfall
Montana	Flowery place
Nevada	Mountain
Palo Alto	The pass
Palomar	Red
Pasadena	Arid zone
Reno	Yellow

16. Why is Yellowstone Park called Yellowstone Park?

17. "No Man's Land," a term we hear in old war movies, was a real name given to a conspicuous portion of an actual American state. Where was it?

18. Here is a list of places, one for each state. Can you fill in the appropriate state?

Arab	Wyoming
Rockcastle	Cucamonga
Wonder Lake	Chloride
Wagon Mound	Hockessin
Coos	Owyhee
Enoree	Hopatcong
Common Fence Point	Mazeppa
Bell Buckle	Venice on the Bay
Ten Sleep	Spearfish
Clintwood	The Village
Gang Mills	Buzzards Bay
Flathead	Black Lick
Yossem's Paradise Valley	Ishpeming
Weed Heights	Black Jack
Dinosaur	Bogue Chitto
Oil Trough	Poteet
Mystic	Fossil
Tok	Candor
Bird City	Tipp City

Battle Ground

Mililani Town

Flowery Branch

Yeehaw Junction

Cut Off

Cape Porpoise

Hoople

Disputanta

Outagamie

Cle Elum

Tropic

Paw Paw

19. Where was the State of Deseret?

20. What do Holy Cross University, the St. Croix River in Wisconsin, and Santa Cruz, California, have in common?

21. What state's name means "owner of big canoes"?

Place Names

1. They originated around Blue Point on the Great South Bay, between Fire Island and Long Island's south shore.

2. Washington.

3. The study of place names.

4. Montana.

5. Of the six New England states, only New Hampshire betrays the English origin of the settlers. Massachusetts and Connecticut are derived from American Indian names. Vermont is a corruption of the French for "green mountain." Maine is a historic county of France. And Rhode Island, at least indirectly, harkens back to the Mediterranean island of Rhodes.

6. These New York cities are all named for classical cities in the Mediterranean region of Europe. Such classic place names are widespread in the United States, which has a Mount Olympus in Washington State.

7. Mission Dolores. It was established as a Spanish mission.

8. Bet you thought it was an Indian name. But H. R. Schoolcraft, something of a pedant, named it in 1832 from the Latin *veritas caput*, meaning "true head."

9. Pekin is in Illinois; Memphis in Tennessee; Babylon in New York; Versailles in Kentucky; Cairo in Illinois; Paris and Palestine in Texas; Moscow in Idaho.

10. Many people of Dempsey's day, recalling the great Civil War battle of Manassas, Virginia, presumed the champ was a Virginian. But Dempsey hailed from little Manassas, Colorado.

11. Wounded Knee, South Dakota; the Finger Lakes district of New York; Cheek, Oklahoma; Braintree, Massachusetts; Finger, Tennessee; Hand, South Carolina; Heartstrong, Colorado; Headville, West Virginia; Throg's Neck, New York; Mouth of Wilson, Virginia; Boca Raton (Rat Mouth), Boca Grande (Big Mouth), Boca Chica (Little Mouth), Florida; Muscle Shoals, Alabama; Organ, New Mexico; Butte, Montana.

12. Hope, Arkansas; Charity, Missouri; Cape Fear, North Carolina.

13. North Carolina, Arkansas, Georgia, West Virginia, New York, Pennsylvania, and Tennessee each has a Fayetteville—there may be others we haven't heard of.

14. These are the Indian meanings of, in order: Wyoming, Chicago, Mississippi, Manhattan, Missouri, Oklahoma, Omaha.

15. Amarillo is yellow, Arizona is dry or arid zone, Colorado is red, El Paso is the pass, Florida is flowery place, Montana is mountain, Nevada is snowfall, Palo Alto is high pole, Palomar is dovecote, Pasadena is strolling place, Reno is reindeer.

16. Yellowstone Park is named for the yellowish volcanic rock called rhyolite, which is a prominent feature there.

17. No Man's Land is what the Oklahoma panhandle was called before statehood. Until 1890, it was outside both federal and Indian law.

18. Places in their states:

Arab, Alabama
Rockcastle, Kentucky
Wonder Lake, Illinois
Wagon Mound, New Mexico
Coos, New Hampshire
Enoree, South Carolina
Common Fence Point, Rhode Island
Bell Buckle, Tennessee
Ten Sleep, Wyoming
Clintwood, Vermont
Gang Mills, New York
Flathead, Montana
Yossem's Paradise Valley, Nebraska
Weed Heights, Nevada
Dinosaur, Colorado
Oil Trough, Arkansas
Mystic, Connecticut
Tok, Alaska
Bird City, Kansas
Battle Ground, Indiana
Mililani Town, Hawaii
Flowery Branch, Georgia
Yeehaw Junction, Florida
Cut Off, Louisiana
Cape Porpoise, Maine
Wyoming, Iowa (or Delaware)
Cucamonga, California
Chloride, Arizona
Hockessin, Delaware
Owyhee, Idaho
Hopatcong, New Jersey
Mazeppa, Minnesota

Venice on the Bay, Maryland
Spearfish, South Dakota
The Village, Oklahoma
Buzzards Bay, Massachusetts
Black Lick, Pennsylvania
Ishpeming, Michigan
Black Jack, Missouri
Bogue Chitto, Mississippi
Poteet, Texas
Fossil, Oregon
Candor, North Carolina
Tipp City, Ohio
Hoople, North Dakota
Disputanta, Virginia
Outagamie, Wisconsin
Cle Elum, Washington
Tropic, Utah
Paw Paw, West Virginia

19. "Deseret" is a word from the Book of Mormon, meaning "honeybee" and signaling the hard work necessary for true success. When the Mormons moved to the Salt Lake Desert, they named their settlement the State of Deseret, but this was not accepted by the U.S. Congress, which instead chose the American Indian name Utah.

20. The same name, more or less. St. Croix and Santa Cruz can both be translated as "holy cross." This is an example of our diverse ethnic heritage and our religious bent.

21. Missouri, in the Iliniwek Indian language.

Men and Women

Do men and women have different personal geographies?

In public places, the sexes generally go to different bathrooms, though they use the same ones at home. There are still typically male and female jobs. Women are believed to solve problems by conciliation while men fight it out.

Do the sexes see their surroundings differently? While there may be no certain way of knowing, it would stand to reason, even in the most egalitarian of societies. Only women can become pregnant—definitely a perception-altering condition. Sex roles and perceptions are certainly changing, and geographers are beginning to study the geography of the sexes. For example, are some parts of cities more female—chosen for their safety? Are some buildings—stadiums, for instance—more frequented by men? Gymnasiums used to be; now there are at least as many women in them, and they're often called spas. Do men and women want different things from a gym? Are there really Norwegian bachelor farmers in Minnesota, as Garrison Keillor describes them (and if so, how do they reproduce themselves)? Here are some questions about the sexes in America.

1. The highest single male-to-female ratio is in which of these?

San Diego _____

New York _____

Denver _____

Miami _____

Minneapolis _____

2. The highest single female-to-male ratio is in which of these cities?

San Diego _____

New York _____

Denver _____

Miami _____

Minneapolis _____

3. Thirty-six percent of men are single. What percentage of women are?

25 percent _____

31 percent _____

40 percent _____

46 percent _____

4. Who is happier single, American men or women?

5. In 1990, women earned an average of 72 cents for every dollar earned by men. How much progress had there been two years later? Women then earned:

70 cents _____

72 cents _____

74 cents _____

76 cents _____

78 cents _____

6. How tall is the average American woman?

5 feet _____

5 feet 1 inch _____

5 feet 2 inches _____

5 feet 3 inches _____

5 feet 4 inches _____

5 feet 5 inches _____

5 feet 6 inches _____

7. As portrayed on American television, 29 percent of the professionals are women. What proportion of professionals in real life are women?

8. Twelve percent of American men are bald or balding at age twenty-five. What percent are bald at thirty-five? At forty-five? At sixty-five?

9. There are 21,000 domestic crimes against women reported each week—usually beatings administered by their boyfriends or husbands. Yet there are three times more _____ shelters in the United States than shelters for battered women. Fill in the blank.

10. Many Americans have awakened to discover they married someone gay or lesbian. About how many are believed to have married a homosexual?

1 million _____

2 million _____

3 million _____

4 million _____

11. Between 15 million and 20 million American women were believed, in the early 1990s, to own what piece of hardware?

12. When we feel anxious, we tend to eat less—or eat more. But men and women do it differently. Who eats more, and who less, under stress?

13. Are there more, or fewer, unmarried Americans since 1970?

14. Six out of ten American unmarried adults between ages twenty and twenty-four live in the same place, in a manner of speaking. Can you guess where?

15. Double marriage question:

(a) At what average age do American men marry?

(b) At what average age do American women marry?

16. Every year, 1.25 million Americans suffer heart attacks. What proportion of them are men?

One fourth _____

One third _____

Two thirds _____

Three fourths _____

17. Who experiences more stress within the family, men or women?

18. Women's bodies are 25 percent fat by weight; what percent fat are men's?

19. In 1988, the American man spent an average of thirty minutes a day on personal grooming. Two years later, how had this changed?

20. How many sperm cells does the average man produce in a month?

Half a million _____

A million and a half _____

About 10 million _____

10 billion to 30 billion _____

21. In 1992, about 800,000 _____ under _____ were reported missing, eight times as many as ten years before. Fill in the blanks.

22. It is commonly known that women tend to live longer than men. But men, if they do survive, succeed better than women in one respect. What is it?

23. Is a man's skin thicker than a woman's skin, or thinner, or about the same?

24. How many American mothers breast-feed their babies?

One fourth _____

One third _____

One half _____

Three fourths _____

25. Nine in ten American women wear shoes at least _____ sizes _____ er than her feet. Fill in the blanks.

26. In 1870, the Census showed about an equal number of men and women in the United States. But they were distributed unevenly. How so, and why?

27. American women who went to college earn an average of $21,700 a year. How much do men who went to college earn?

    $25,600 _____

    $31,400 _____

    $36,000 _____

28. What percentage of the characters on prime-time TV are male?

29. Two-part diet question:

    (a) On an average day, how many American women are on a diet?

    About 10 million _____

    About 20 million _____

    About 40 million _____

    About 50 million _____

    About 60 million _____

    (b) How many men?

    About 10 million _____

    About 20 million _____

    About 40 million _____

    About 50 million _____

    About 60 million _____

30. What proportion of American men and women are within their recommended body-weight range?

31. Seventy-eight percent of American women wear colored underwear. What percentage of men do?

    About 4 percent _____

    About 14 percent _____

    About 24 percent _____

32. Without training, a man's "natural" strength increases from adolescence till age twenty-five or thirty. A woman's natural strength remains at the level attained at age:

    Ten _____

    Twelve _____

    Twenty-two _____

    Thirty-two _____

    Thirty-eight _____

33. How many American men think hunting and killing animals for sport is acceptable?

    About 90 percent _____

    About 50 percent _____

    About 22 percent _____

34. Forty-nine percent of the homeless are single males. Who are the rest?

35. What proportion of American women say the amount of money their intended mate earns is important to them?

    About 21 percent _____

    About 35 percent _____

    About 58 percent _____

36. What is the record for most monogamous marriages in the United States?

    By a woman _____

    By a man _____

37. For every 100 baby girls born, how many baby boys are born in the United States?

38. Is a man, or a woman, more likely to turn down a heart transplant even if it is the only hope to save his or her life?

39. In 1916, when most women in the United States were not permitted to vote, Montana did something very unusual. What was it?

40. You know that feeling of infatuation—when you've fallen for someone and are walking about an inch off the ground? By and large, how long does it last among American men and women?

41. Evolution, they say, takes place partly by genetic mutation. In humans, one gender produces more genetic mutations than the other. Which one?

42. Is the psychological advantage of being married greater for men or for women?

43. Of a particular intimate activity between men and women, Nevada has the highest rate in the United States, while Pennsylvania has the lowest. Can you name the activity?

44. In 1970, 6 percent of the women and 10 percent of the men reaching age thirty were still unmarried. Two decades later, how would you say those figures had changed?

45. Many thoughtful people have noticed that men tend to be more violent than women. Is testosterone the reason?

46. About how many American women wear at least a size 12 dress?

47. About how many American men don't have a best friend?

Men and Women

1. San Diego, because of its enormous naval base.

2. New York, who knows why?

3. Forty percent of American women are single.

4. Women, especially if they have enough money. The old stereotype of the woman panting to find a mate is completely wrong nowadays. Financially and sexually independent women increasingly regard marriage as a poor bargain. One sociologist calls this attitude "the real revolution" of the past twenty years. "They can support themselves in reasonable style," said Frances Goldscheider of Brown University. "They don't define themselves around men. And they do well socially because they have friends and bonding skills." American men, who have often depended upon women for their social connections, may do less well. They may want to remarry quickly after a divorce, to have their needs fulfilled, but finding a willing woman may be difficult. "It's hard for men to remarry when women aren't interested," said Barbara Foley Wilson, a demographer at the National Center for Health Statistics.

5. Women's earnings were down 2 cents to the male dollar in 1992. Younger women, however, those aged twenty-four to thirty-five who came to work with better skills and education, earned 80 cents to every dollar earned by men the same age. (The highest-paid woman in the country in 1992 was Turi Josefsen, executive vice president of U.S. Surgical Corp. She was earning $23.6 million a year.)

6. The average American woman is 5 feet 4 inches.

7. About 50 percent of professionals in real-life America are women, many of them teachers.

8. At thirty-five, 37 percent of men are bald; at forty-five, 45 percent are bald; and at sixty-five, 65 percent are bald.

9. There are three times more animal shelters.

10. About 2 million Americans are thought to have married someone who later either discovered or acknowledged himself or herself to be homosexual. Many divorce, and some turn their relationship into a strong platonic friendship.

11. This many American women are armed with guns.

12. On average, men eat less when they experience anxiety; women eat more.

13. The number of unmarried Americans doubled between 1970 and 1990.

14. With their parents.

15. The average man marries at twenty-six and a half; the average woman at twenty-four and a half.

16. Two thirds of heart-attack victims are men.

17. Women report stress in family relationships much more often than men.

18. Men's bodies are 15 percent fat.

19. In 1990, men groomed themselves forty-four minutes a day. Extra credit: Ponder how such changes, which appear to be private, individual decisions, actually occur across the entire culture at the same time.

20. The average man produces 10 billion to 30 billion sperm a month.

21. Children under eighteen. Many run away. Some are taken by a family member, some are kidnapped for money or molestation. About 150 a year are known to be murdered after kidnapping. Many more are never found. About 200 to 300 fit the frightening stereotype of kidnap by a total stranger and are murdered or held for ransom. Of the 350,000 abducted yearly by a family member, 1 percent or 3,500 are not returned.

22. If an elderly man survives, he is more likely than a woman his age to be healthy, self-sufficient and living with his family rather than alone.

23. Male skin is thicker.

24. About half of American mothers breast-feed.

25. Two sizes narrower.

26. There were more men—especially young men—than women in the West and Midwest, while more women than men remained in the East. This reflected exploratory patterns.

27. Men who went to college earn an average of $36,000, nearly twice what college-bred women earn.

28. Sixty-seven percent are male.

29. On a given day, 60 million American women and 41 million American men are on a diet.

30. About 24 percent of women and 22 percent of men are within the recommended weight range. That means more than three fourths of us are either overweight or underweight, most likely over, and that more men need to lose weight than women. We're all kidding ourselves, too. In one study, women overestimate their overall body measurements by 25 percent while men overestimate by only 13 percent.

31. About a fourth of American men wear colored underwear.

32. A woman's natural strength, without training, remains about what it was when she was twelve.

33. About half of American men think it's acceptable to hunt and kill animals; half don't. Among women, only 24 percent think it is acceptable.

34. Thirteen percent are single females; the rest are married people, some with children.

35. Money is an important mating consideration for 58 percent of American women who admit it.

36. The woman's record is sixteen; the man's record is twenty-seven.

37. One hundred five boys are born for each 100 girls. But over time, the rate changes. By the time they are eighteen, the numbers even out—100 males to 100 females. By age sixty-five, there are only 68 males to each 100 females, and by age eighty-five, there are only 44. The ratio of men to women in the United States overall is 95 to 100.

38. Women are more likely to reject the opportunity for a heart transplant. Researchers suspect they are more accepting of death's inevitability.

39. Montana sent Jeannette Rankin, the country's first congresswoman, to Washington, D.C.

40. The duration of romantic love in America is between eighteen months and three years, according to psychologist Dorothy Tennov. She measured the period from when infatuation first struck to the time when a "feeling of neutrality" set in. Some other researchers believe the end of infatuation is inevitable, rooted in brain chemistry and physiology. It's too draining to sustain.

41. Men are responsible for six genetic mutations for every one caused by women. This is because men produce more sperm than women produce eggs and thus have more chances to make a "mistake."

42. The advantage is greater for men. A Yale study found that, among husbands and wives with no history of depression, divorcing produced three times more cases of depression among the men.

43. Marriage. People from all over the world go to Nevada to divorce and marry. Pennsylvania has the lowest rate because it has the fewest people in the marrying ages.

44. Those percentages approximately tripled in two decades. About 19 percent of women, and nearly 30 percent of men, remained unmarried at thirty.

45. One convincing study suggests the cause is not genes or hormones, but cultural attitudes. "Both sexes see an intimate connection between aggression and control," says Anne Campbell, of the School of Health, Social and Policy Studies of Teesside University in England, who studied both Britons and Americans, "but for women aggression is the *failure* of self-control, while for men it is the *imposing* of control over others. Women's aggression emerges from their inability to check the disruptive and frightening forces of their own anger. For men, it is a legitimate means of assuming authority over the disruptive and frightening forces in the world around them," which, very often, seems to include women.

46. More than half the women in America wear at least a size 12.

47. More than half the men in America don't have a best friend, according to Dr. Alvin Baraff, director of MenCenter in Washington, D.C. "The guy who says he has more than two good friends is a most unusual person who either has a special knack or is exaggerating."

Family

Though the family is an icon in the social geography of America, professional geographers have given very little attention to it. We may see this changing as the traditional family unit is taken less for granted, which appears to be happening.

Many of us still get together at holiday time, or at great gatherings called family reunions, at which we rediscover how much we resemble each other, how frequently we tend to have the same habits and diseases. But there is little doubt that our assumptions about family are getting reexamined. Adoptees now routinely knock on their biological parents' doors. Court dockets are filled with cases that would have been impossible a few years ago: surrogate motherhood disputes, gay and lesbian couples, test-tube babies, ownership of frozen sperm or egg after divorce, parental responsibility for accidental shootings of and by children.

Family "problems" are often attributed to certain places and regions. We speak of incomplete families in the inner cities, yet everyone knows of poisonously dysfunctional families in the suburbs, and complete, loving ones in town. Despite statistics and a few clear trends, predictions are still impossible.

1. Mom at home with two children, Dad out working as the breadwinner. This is a portrait of an American family. What proportion of American families actually fit this picture?

2. Where was the long-running television series *All in the Family* set?

3. One of America's most famous radio families was portrayed on the program *One Man's Family*, whose probably sexist title didn't discourage listeners in the 1950s and 1960s. Where was it set?

4. One in 300 teenage American girls has a sad, sometimes tragic delusion. What is it?

5. Who is more likely to divorce in America: poor couples or rich couples?

6. Among American teenagers aged thirteen to seventeen, what proportion say they have had sex?

7. What proportion of Americans say they would donate a deceased family member's organs to a stranger who needed them without that member's prior permission?

8. In 1970, about 10 percent of American families were headed by single mothers. By 1991, how had that figure changed?

9. One new style of American family, a household of older women, was portrayed in *Golden Girls*. Where was it set?

10. Between 1980 and 1990, the number of single fathers living with children more than doubled in the United States. By the early 1990s, fathers were heading what percentage of single-parent households?

    2 percent _____

    4 percent _____

    11 percent _____

    14 percent _____

    33 percent _____

11. What proportion of American homes had personal computers in 1994?

12. In 1940, nine out of ten households included a married couple. What proportion of households did by the early 1990s?

13. Family ties are breaking down in the United States, correct? We devour our young and abandon our elderly?

14. One of America's odder TV families was the Bundys, portrayed on *Married with Children*. Where was the show set?

15. Who are Beldar and Prymaat?

16. "It's an astonishing development," the famous Harvard social scientist David Riesman said in mid-1993. It is "an enormous change . . . with profound effects on the quality of life in this country." Of what social change was Riesman speaking?

17. The number of people over eighty-five will grow from 3.3 million in the early 1990s to 18.7 million in the year 2080, if you believe the Census Bureau. If you believe the National Institute on Aging (NIA), the number could be nearly four times greater, or 72 million. What accounts for the difference in these two estimates?

18. The number of interracial marriages in America doubled between 1980 and 1992. What proportion of couples is interracial?

19. Double ring query:

(a) What percentage of married men wear a wedding ring?

(b) What percentage of married women?

20. During an argument with her husband, a wife's blood pressure increases an average of 6 percent. How much will a husband's blood pressure rise during the same argument?

About 1 percent _____

About 3 percent _____

About 6 percent _____

About 12 percent _____

About 14 percent _____

21. What percent of working mothers say they prepare dinner alone?

About 55 percent _____

About 66 percent _____

About 77 percent _____

Family

1. Only 5 percent of American families fit the idealized picture.

2. *All in the Family* was set in Queens, New York.

3. *One Man's Family* was set in San Francisco.

4. One in 300 teenage girls believes, though she is normal or thin, that she is fat, and suffers chronic appetite loss as part of her effort to "correct" the problem. Some starve themselves to death as a result of their anorexia.

5. "It appears that stresses arising from low income and poverty may have contributed substantially," says Donald Hernandez, author of a recent study which found that one of seven couples below the poverty line splits up, while only one in thirteen with higher incomes does. When the better-off couple splits, three times out of four the mother—with her children—then falls below the poverty line. "Strong family values have to include a strong economic foundation for families," says Clifford Johnson of the Children's Defense Fund.

6. In 1988, 33 percent said they had had sex; 72 percent of the boys (versus 56 percent in 1979), and 27 percent of the girls, up from 19 percent in 1982. Forty-eight percent said they felt sex before marriage was wrong.

7. Just 47 percent would donate the organ to a person who needed it.

8. The number of single mothers had more than doubled, to 21.1 percent, and was still rising.

9. *Golden Girls* was set in Miami.

10. Single fathers headed 14 percent of single-parent households.

11. About one in three homes had one or more personal computers—about 30 million of them.

12. In 1990, the most common household in America was two persons living together. Only 55 percent of American households included a married couple. The number of households has been growing—to 95.7 million in early 1993, up 2.3 million in three years. One in four was a person living alone.

13. No, not correct at all—or certainly not entirely. Despite the tales we've all heard, an exhaustive federal study in 1993 found that middle-aged Americans take good care of their elderly parents. In addition, 30 percent to 40 percent of Americans in their fifties with children were helping the children financially and in other ways. At least 60 percent of Americans over eighteen had parents either living with them or within an hour's drive. And all those jokes about old parents who never heard from their offspring were also belied. Some 85 percent saw or spoke to their children two to seven times a week.

14. Chicago.

15. Beldar and Prymaat are the elder Coneheads, just a couple of kooks from the planet Remulak, not far from Krypton. Played by Dan Aykroyd and Jane Curtin, the 'heads appeared on *Saturday Night Live* television skits and in a feature film. They love pizza, which they call "starch disk."

16. Riesman was speaking of the fact that, while only 26 percent of Americans were single two decades

ago, nearly 40 percent are today, and the number is growing. The change is driving everything from novel political agendas to the invention of the one-person coffeepot. Riesman, author of the 1950 classic study *The Lonely Crowd,* said, "What is a society like when individualism so often expresses itself on a computer screen?" This signals, he says, an even more lonely, isolated, selfish, and self-absorbed society. Other observers think the spreading of singleness is behind the growth of confessional television talk shows that provide "surrogate intimacy, a substitute for the sense of connection that the family used to provide." Marriage may be less common but, ironically, the birth rate is up 10 percent since the 1970s.

17. The difference is in whether one believes there is a "natural limit" to the human life span, or that there is no built-in biological cap on longevity. We are a healthier country in general, and old people are entering the high numbers of old age in much better shape than previously. Dr. James Vaupel, a Duke University demographer who produced the NIA estimate, says there's no reason today's children cannot routinely live to be 100 or more. Other scientists note that longevity gains become harder and harder as the ages get higher and higher. It is still true that the older one is, the more likely one is to die. But many scientists are saying, as does one: "There is no fixed limit to the human life span. It could be anything."

18. One in fifty couples in America is interracial.

19. Seventy-six percent of married men wear a ring, but only 56 percent of women do.

20. A husband's blood pressure rises about 14 percent during a domestic spat.

21. About 77 percent of working mothers say they come home from the job and cook dinner alone. Only 16 percent say their husband does. And 64 percent say they clean up after dinner without male help. Well, gee, there was probably a game on the tube.

Ethnic Stew

Amerindians came first. Then came others from every direction, for every conceivable reason, to settle in America. Africans came unwillingly. Few countries are as ethnically diverse as the United States, although Russia, India, and Brazil may come close.

We pay lip service to the idea that our diversity is a strength, while in practice we have been as prejudiced and xenophobic as any other humans. Perhaps something in our ancient makeup causes us to categorize others, and maybe what distinguishes those who live together successfully—with empathy, cooperation, and humor—is how well they handle this impulse. Comfortable economic times, of course, make it easier.

There are certainly great advantages to living in an ethnically diverse country, or part of the country. Life is more interesting. You can find dark-roast coffee, salsa picante, bok choy, kielbasa, kimchee, or cuttlefish marinara when you want some. If you make a stew—or a goulash, bouillabaisse, ragout, paella—you want the flavors to marry but not merge. So far in America, the melting pot is less a fact than a beautiful idea, one that has etched its marks deeply into our country's landscape.

1. Which ethnic ancestry is the most common among Americans?

English _____

Irish _____

German _____

Italian _____

African _____

Polish _____

2. Which city has the country's highest percentage of foreign-born residents?

Los Angeles _____

Miami _____

New York _____

Hialeah _____

Santa Fe _____

Union City _____

3. One ethnic minority will double in population by 2020, at which time it will represent 6 percent of Americans. Which is it?

4. In 1966, African-Americans comprised 1.7 percent of the nation's professionals: doctors, lawyers, etc. By the early 1990s, they were:

1 percent _____

5 percent _____

9.6 percent _____

12 percent _____

22.2 percent _____

5. Hate crime grew in the United States during the 1980s' atmosphere of us-against-them. New York and New Jersey led the nation in the number of hate crimes, according to an FBI study, although some states did not provide statistics. What would you say was the main motive in most hate crimes?

Race _____

Religion _____

Gender _____

Sexual orientation _____

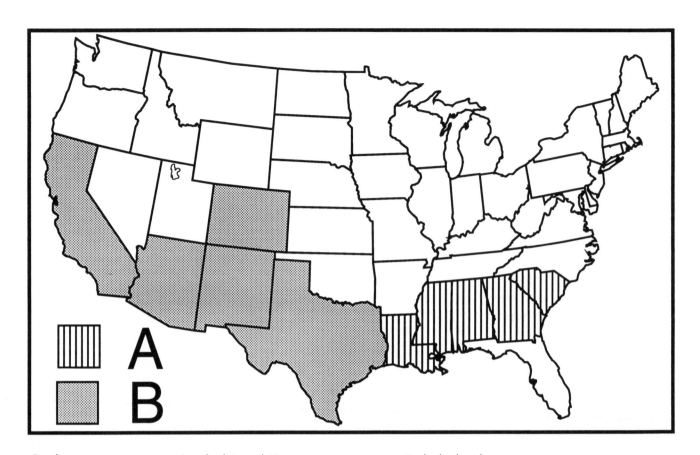

On this map, two patterns (marked A and B) appear on ten states. Both deal with the human geography of the United States. What do patterns A and B demarcate?

Answer: Pattern A denotes the five states with the highest percentage of black residents. Mississippi is highest, with 36 percent. Pattern B denotes the five states with the highest percentage of Hispanic residents. New Mexico is highest with 38 percent Hispanic. However, the most black Americans live in New York, California, and Texas (in that order), and the most Hispanics live in California, Texas, and New York (in that order).

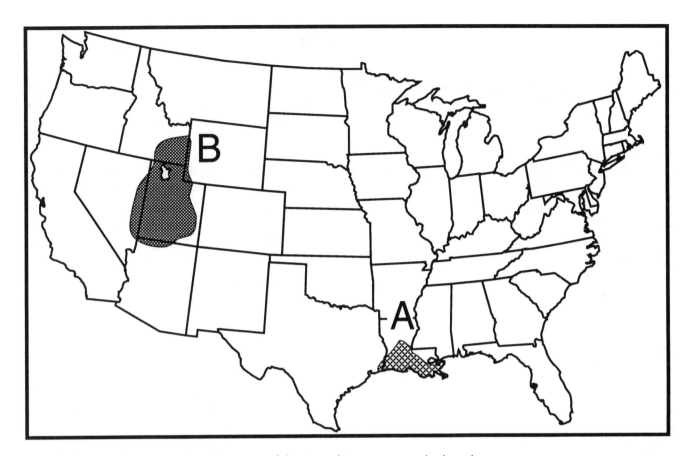

Two of the most distinctive cultural regions of the United States are marked on the map above. They are as different from one another as they are from the rest of the country. What are the cultures of region A and region B?

Answer: Geographers call Region A Louisiana French, but the residents often call themselves Cajuns. The word is a slurred version of "Acadian," which is an old name for their ancestral home, Acadia, now known as Nova Scotia. This region has everything from andouille to zydeco, and bottles more versions of hot sauce than Scotland does of single-malt whisky. Region B is the heartland of the Mormons—the Church of Latter Day Saints.

6. In racially motivated hate crimes, one race (see above) was singled out for attack most often. Which race was attacked second-most?

7. What do these urban areas have in common?

Manhattan

Queens

Brooklyn

The Bronx

Los Angeles

San Francisco

Oakland

Jersey City

Greater Houston

8. How many Americans are anti-Semitic?

1 in 100 _____

1 in 50 _____

1 in 20 _____

1 in 10 _____

1 in 5 _____

9. What proportion of the American population is foreign born?

10. Why do many white American parents claim to believe that mathematical ability is inborn?

11. What proportion of American babies and children awaiting adoption are black?

12. Why do some experts now say we should address AIDS partly as a racial issue?

13. What is the most common cause of death among African-American teenagers?

14. Which American region is the most stereotyped, caricatured and jeered at? (Hint: It was the home of the Double Bubba Ticket.)

15. Which state had the largest number of black elected officials in the early 1990s?

Alabama _____

Missouri _____

Minnesota _____

California _____

New York _____

Illinois _____

16. How many blond(e)s are there on television compared to the number in real life?

Half as many _____

About the same _____

Twice as many _____

17. What proportion of American-born children are biracial?

About 12 percent _____

About 22 percent _____

About 32 percent _____

About 41 percent _____

18. On the average, white Americans feel comfortable standing about 27 inches apart while conversing. About how far apart do black Americans prefer to stand?

About 22 inches _____

About 27 inches _____

About 34 inches _____

19. Who was the first African-American to visit what became the United States?

20. Match these Indian groups with their modern home states:

Timucua	Alabama
Shawnee	Texas
Miami	New York
Tillamook	Montana
Costanoan	Washington
Yakima	California
Crow	Oregon
Mohawk	Indiana
Comanche	Ohio
Alabama	Florida

21. From which part of Africa, generally, did the slaves sold in the United States come?

22. What state has the highest proportion of "Native Americans"?

23. What state has the highest proportion of people of Japanese descent?

24. Where was the first French capital of the "Mississippi Valley"?

25. In 1900, the life expectancy for white Americans was forty-four years. What would you say it was for black Americans?

26. In South Africa in the early 1990s, 681 black men per 100,000 were in jail. What would you say the comparable figure was in the United States?

About 50 _____

About 500 _____

About 1,000 _____

About 2,000 _____

About 3,000 _____

27. What ethnic group in the United States controls 56–58 percent of the heroin trade?

African-American _____

Vietnamese _____

Italian _____

Chinese _____

French _____

Turkish _____

28. About how many historically black colleges are there in the United States?

29. In the 1990 census, Americans were given four basic choices to identify themselves by race: White; Black; American Indian or Alaskan Native; Asian or Pacific Islander. A fifth category was "Other." Who were the 10 million Americans who did not find themselves in the first four categories and termed themselves as "other"?

30. There were 30,432 postdoctoral fellows in science and engineering in the United States in 1991, according to the National Science Foundation. About how many of them were foreign students?

31. Hawaii has the greatest proportion of Native Americans, but what state has the *most* Native Americans?

32. What state has more ethnic Poles than any other?

California _____

Florida _____

Illinois _____

New York _____

Michigan _____

33. Match what state has the most citizens of:

Cuban descent   New York

Mexican descent   California

Puerto Rican descent   Florida

34. Which state has the most black people?

35. Which state has the most residents of Hispanic origin?

Ethnic Stew

1. In the 1990 U.S. Census, 23 percent of Americans (nearly 58 million) claim German ancestry, followed by 15.6 percent Irish, 13.1 percent English, 9.6 percent African-American (others claimed "mixed" race; final census figures say 12 percent are African-American), 5.9 percent Italian, 5.0 percent American, 4.7 percent Mexican, 4.1 percent French, 3.8 percent Polish, and 3.5 percent American Indian.

2. Hialeah, Florida, a part of the Miami megalopolis, has 70.4 percent foreign-born residents, making it number one in the nation. Miami is number two.

3. Asian-Americans.

4. Blacks were 5 percent of the professional classes. But since they account for some 12 percent of the population, they were still underrepresented. They also remained underrepresented as bartenders (only 2 percent are black), dental hygienists (1.1 percent) and speech therapists (1.3 percent). But they had plenty of representation in such less desirable jobs as janitor (21 percent), short-order cook (23.3 percent), and nursing aide (31.2 percent).

5. Race was the most common factor in hate crimes. Sixty-two percent of the incidents were race-related, with blacks the main target. Fifty-seven percent of hate crimes were against African-Americans. Among religion-related incidents, 86 percent were against Jews.

6. Whites.

7. These are the most ethnically diverse cities in the country. Their populations are evenly balanced between non-Hispanic whites, non-Hispanic blacks, Hispanics, and non-Hispanic other races.

8. Anti-Semitism has declined in America since the sixties, but nearly 40 million of us, or one in five, still believe that "Jews have too much power in the business world." Americans were most likely to be anti-Semitic if they were over sixty-five, or poorly educated, or blue-collar, or black, or all of the above. On college campuses, anti-Semitic incidents have increased, especially among African-Americans. "Many black student leaders and representatives," said an Anti-Defamation League report, "repeatedly and enthusiastically support speakers who are well known for their Jew baiting."

9. Eight percent of the American population—or 19.8 million people—is foreign born. The most common country of origin today is Mexico, with 21.7 percent of the total, followed by the Philippines (4.6 percent), Canada (3.8 percent), Cuba (3.7 percent), Germany (3.6 percent), Britain (3.2 percent), Italy (2.9 percent), South Korea (2.9 percent), Vietnam (2.7 percent), and China (2.7 percent).

10. Probably to excuse the poor showing their children make on standardized math tests in comparison to Japanese and Chinese students. However, they do not claim that "general knowledge" is inborn, since American kids do better in that culturally loaded area.

11. The best estimates in the early 1990s suggested 40 percent of children hoping to be adopted were black. Because African-Americans are only some 12 percent of the population, many of these chil-

dren were doomed to lives in foster care unless they were adopted by persons of other races. Yet transracial adoption has been discouraged in America, and only accounted for about 12 percent of all adoptions. Most of these have been whites adopting blacks for idealistic and religious reasons.

12. Nearly half the people with AIDS in the United States are black or Hispanic, though these groups account for only 21 percent of the population. But it isn't ethnicity that matters, it's money. "Low income and poor health are strongly linked," said a report by the National Commission on AIDS. "Poor people of color often are isolated from all but the most rudimentary health care." It would make sense, then, to target them as a high-risk group.

13. Homicide kills more black teenagers than any other cause, at a rate nine times higher than for white teenagers. But, when you combine the statistics for suicide, accident, and homicide, young blacks die at a rate only 1.4 times that of whites. Still higher, but closer, and some observers suggest that, as the scholar Andrew Hacker put it, "while despair remains more widespread among black youths—which is to be expected—a not dissimilar despondency may be found among whites."

14. The South, of course, is the most caricatured of all United States regions. Any ethnic group that was laughed at as much as white Southerners have been would have long ago lost their sense of humor. A 1990 study of ethnic images found other Americans see white Southerners as a kind of ethnic group all its own, "a little less intelligent, a little less hardworking" than whites in general. "I don't believe that non-Southerners' repertoire of stereotypes really has space for middle-class Southern folks," said one white Southerner.

15. Alabama had the most black elected officials (706), followed by Mississippi (690), and Louisiana (551).

16. Proportionately, there are twice as many blond(e)s on TV as in real life.

17. Thirty-two percent of American-born children are biracial.

18. On average, black Americans prefer a 22-inch conversation distance.

19. A freed slave named Esteban was part of a Spanish expedition to the Tampa Bay area of Florida in 1536. Esteban later led explorers in the southwestern region of what became the United States. But more likely, this distinction belongs to some unknown African who may have escaped slavery in Cuba and crossed what became the Florida Straits, just as Cuban refugees do today.

20. The Timucua lived in what is now Florida, the Shawnee in Ohio, the Miami in Indiana, the Tillamook in Oregon, the Costanoan in California, the Yakima in Washington, the Crow in Montana, the Mohawk in New York, the Comanche in Texas, and the Alabama in, you guessed it, Alabama.

21. Most U.S.-bound slaves were shipped from the coast of what is now Nigeria, Ghana, and the Ivory Coast, and they lived in the large region inland of this coast. They were placed in slave ships (many of which had holds divided for men, girls, and boys), and shipped across the ocean like cattle to the New World. Only about 6 percent of enslaved Africans were sent to the actual territory of the United States.

22. Hawaii has the highest proportion of "Native Americans."

23. Hawaii has the highest proportion of Japanese-Americans.

24. Biloxi, Mississippi, founded in 1699. The capital was moved to New Orleans in 1722.

25. The life expectancy for a black American in 1900 was about thirty-three years.

26. In the United States, 3,370 black men per 100,000 were incarcerated.

27. Ethnic Chinese triads control most of the American heroin trade.

28. There are 107 historically black colleges in the United States, with about 270,000 students. One eighth of those students are white, and the proportion is growing by about 5 percent every year.

29. Most were found to be Hispanics. "Hispanic" is considered to be an ethnic category, under "Spanish surname," but has not been regarded as a racial one since 1930 by the Census. Some odd answers turned up on the census forms. In 1990, for instance, 1.8 million identified themselves as American Indians, and then 8 percent of these went on to claim membership in the following tribes: Polish, Haitian, Hispanic, Arab, African-American.

30. About half the science and engineering postdocs were foreign students, up from a third ten years before. And a third of U.S. doctoral recipients were foreign students—the majority also from Asia.

31. Oklahoma, in 1990, had 252,000 Indians, more than any other state.

32. New York has the most ethnic Poles.

33. This was easy: Cuban, Florida; Mexican, California; Puerto Rican, New York.

34. New York state had nearly 3 million who, in the 1990 census, identified themselves as black. The state with the highest percentage of blacks is Mississippi, with 36 percent.

35. California has the most Hispanics—nearly 7.7 million in 1990. The state with the highest proportion of Hispanics, however, is New Mexico, where they are 38 percent of the population. According to the U.S. Census, you belong to whatever racial or ethnic category you choose, no matter your physical attributes.

Migration

Very often in life, you will have noticed, what you want turns out to be somewhere you're not. Before mail-order catalogs, the only way to get the object of your desire was to trade for it or go where it was—to migrate. Humans have done this since Day One, whenever that was.

Usually, migration has "push" elements as well as "pull" elements. Some places are simply more pleasant than others to live in, and thus, all other things being equal, will attract more migrants. Others drive people away. Today, for instance, the fastest-growing part of America is the Sunbelt, a region embodying many of the reasons people migrate. They come from the north for sunshine, fun, and an active economy in which they are more likely to find work. They come from parts of Latin America, to the south, for jobs and to avoid being tortured and shot by rebels or their government.

As people migrate, they bring with them ideas of how a landscape should look, plants they like to eat and look at, animals they feel comfortable with as pets and food. They bring their own music, clothing, religions, and cuisines. Before long, they have modified the new place—yet without quite managing to turn it into a replica of the place they left.

1. "The great social adventure of America is no longer the conquest of the wilderness, but the absorption of fifty different _____ ." Fill in the word missing from this quotation from Walter Lippmann.

2. One state has traditionally been a golden destination for Americans on the move, but more are now leaving that state than migrating to it. Name the state.

3. True or false: One in three persons moving into California in 1992 was an immigrant.

4. Which is the fastest-growing state?

5. The year is 1650. How many people of European descent are living in the British colonies of eastern North America?

   About 10,000 _____

   About 25,000 _____

   About 50,000 _____

   About 200,000 _____

   About 500,000 _____

6. Where was the "Trail of Tears," who traveled it, and how did it get its name?

7. Why do some Texans sound a lot like Tennesseans?

8. The peak year of Irish immigration to the United States was 1851—ten years after what well-known event?

9. There were two peaks in German immigration to the United States. The first was in 1882. The second followed what well-known event?

10. In 1890, the U.S. Census Bureau officially declared that the frontier was closed. Nineteen years later, something happened to American immigration as a result. What was it?

11. In 1920, about 35 percent of the citizens of Maine and Vermont had been born in a foreign country. What country was this?

12. What period during the twentieth century would you say had the lowest number of immigrants to the United States?

13. About how many illegal aliens were in the United States in the early 1990s?

14. Since 1930, two great patterns of internal migration have influenced American life in everything from income to literature to popular culture. What are these?

15. Which was first—the pilgrims' arrival in Massachusetts, or the founding of Santa Fe, capital of the Spanish colony of New Mexico?

16. The first group of Jews in the United States arrived in New Amsterdam in 1654. From where did they come?

17. Which Asian country has sent the United States the most immigrants since 1961?

Thailand _____

Vietnam _____

Cambodia _____

The Philippines _____

China _____

18. Which European country has sent us the most immigrants since 1961?

The United Kingdom _____

Greece _____

France _____

Germany _____

Yugoslavia _____

Romania _____

19. Which North American country sent us the most immigrants since 1961?

20. What South American country has sent us the most immigrants since 1961?

Colombia _____

Brazil _____

Argentina _____

Chile _____

Ecuador _____

21. What African country has sent us the most immigrants since 1961?

South Africa _____

Egypt _____

Sudan _____

Cameroon _____

Zaire _____

22. From what European country do the most foreign visitors to the United States come?

23. From which country in North America do most foreign visitors to the United States come?

24. From which in Asia?

25. From which in Africa?

26. From which in South America?

27. What state gets the most foreign immigrants?

28. In what foreign country do the most American citizens live?

29. In 1989, which foreign country was entertaining the most American tourists?

30. In the most recent year for which figures are available, 46.5 million foreign _____ s were expected to visit the United States. Fill in the blank.

31. The first twenty black slaves entered North America in 1619. The number grew slowly over the next thirty years, but then swelled much more rapidly, with results we see around us today. What caused the numbers to begin growing so much faster?

32. What is the origin of the phrase "sold down the river"?

33. What do these descriptions represent:

From east to west

From south to north

From north to south and southwest

From rural to city

From city to city

34. Half a century ago, most immigrants to the United States routinely sought to become citizens. Today, what proportion seeks citizenship?

35. A modern American Indian, Adam Nordwall, a leader of the Chippewa tribe, made a great geographical discovery in 1973. Do you know what it was?

36. What single law induced Americans to settle the Great Plains?

37. Between 1985 and 1990, which state had the most people move in?

38. Who were "The Virginians" who canoed on the River Thames in London, England, in 1603?

39. Match these well-known Americans with the country of their birth:

John James Audubon	England
Alexander Graham Bell	Germany
Irving Berlin	Scotland
Andrew Carnegie	Scotland
Albert Einstein	Russia
Alfred Hitchcock	Italy
Henry Kissinger	Norway
John Muir	Haiti
Joseph Pulitzer	Germany
Rudolph Valentino	Scotland
Knute Rockne	Hungary

Migration

1. Peoples.

2. California. Many residents are moving north to Washington, Oregon, and Idaho. Even so, the population grew 600,000 in 1991, mostly through childbirth and immigration from Mexico and Asia.

3. True.

4. Nevada is growing fastest, followed by Alaska. In both cases the populations are so small that even small numerical increases amount to a large percentage rise.

5. Only about 50,000 lived in all of eastern North America. They never once had trouble finding a parking space.

6. The Trail of Tears refers to the forced removal, in the 1830s, of the Cherokee Indians from the southern Appalachians of North Carolina and Tennessee to what became Oklahoma. Of the 16,000 who began the journey, 4,000 perished en route. This particular trail of tears enjoys the distinction of capital letters, but there were many others in which native peoples were removed from lands the Europeans wanted.

7. Much of northeastern Texas was settled by Tennesseans, both before and after the Civil War.

8. The Irish potato famine.

9. The second German "invasion" of the United States followed World War II.

10. This year, 1908, was the year of most return flow of immigrants to Europe. With the frontier closed, and the Industrial Revolution creating many jobs in Europe, the United States and Europe had begun to resemble each other more and more closely, a process which continues into the present.

11. Canada. Canadian connections with the United States were established in the American Revolution and remain close.

12. Immigration to the United States slacked off during the Great Depression of the 1930s. In fact, in 1932, more people left America than entered.

13. About 3 million illegal aliens is the best estimate. Of course, no one knows for sure.

14. First, the migration from east to west, as ambitious Americans headed for the Pacific shore and the Southwest. And second, the migration of poor Americans, black and white, from the rural, agricultural south to opportunity in the industrial North and Midwest.

15. Santa Fe was officially founded in 1609; the Pilgrims arrived eleven years later.

16. Twenty-four Sephardics, led by Jacob Barsimon, were fleeing the Inquisition—in Brazil.

17. The Philippines has sent most, a total of more than 800,000.

18. The United Kingdom has sent most, a total of more than 400,000.

19. Mexico has sent the most, a total of more than 2,000,000.

20. Colombia has sent the most, a total of more than 240,000.

21. Egypt has sent the most, with a total of more than 70,000.

22. The United Kingdom. Germany is second.

23. Mexico.

24. Japan.

25. Egypt, with Nigeria second.

26. Brazil, with Venezuela second.

27. California gets nearly half of all arriving immigrants. Greater Los Angeles alone accepted nearly a third of all immigrants arriving in 1989.

28. More live in Mexico than anywhere else—a total of 425,000 in 1989.

29. Canada is the foreign country we visit most—390,000, for instance, on May 1, 1989. It's only fair that we are also the country Canadians visit most.

30. Tourists.

31. Production of cotton, rice, indigo, and tobacco had accelerated to meet demands of the European markets. Slaves were more efficient than white indentured servants, who had to be treated somewhat decently and eventually freed.

32. It was a threat to uncooperative slaves who, if they did not behave, would be sold down the river to Mississippi, known as the harshest slave state.

33. These describe the major migratory patterns of Americans over the first half of this century. On top of these have come the enormous immigration of Spanish-speaking persons from Cuba, Mexico, Central and South America, and even more recently, Asians from southeast Asia and China.

34. Only a few more than a third of immigrants seek U.S. citizenship. The other 10 million new residents pay taxes, but they don't vote or serve on juries or become police officers. They dream, most say, of returning to their home countries.

35. Nordwall discovered Italy. He got off an Alitalia jet at Fiumicino Airport in Rome, wearing full Chippewa apparel, and promptly claimed Italy for the Indian people "by right of discovery." Of course, the Italians didn't think they needed discovering, but neither did the Indians when Columbus showed up in 1492, this being the witty Nordwall's point.

36. The Homestead Act of 1862, which entitled a citizen, or an immigrant who intended to become one, to 160 acres. As a result, said one historian, "A larger domain was settled in the last three decades of the century than in all America's past."

37. California, with 3,473,441, followed by Florida with about a million less.

38. They were Indians brought to England by English explorers. Many American Indians were either persuaded, coerced, or dragged to Europe in the 1500s and 1600s, where they became a public sensation and some were presented at court. A few stayed on in Europe.

39. Audubon was born in Haiti, Bell in Scotland, Berlin in Russia, Carnegie in Scotland, Einstein in Germany, Hitchcock in England, Kissinger in Germany, Muir in Scotland, Pulitzer in Hungary, Valentino in Italy, Rockne in Norway.

Language

Language is how we communicate. Need we say more? Well, perhaps. Geographers look at language in terms of who says what where. But language is one of humanity's most portable creations. It spreads almost by itself. The French worry enormously about the pollution of their lovely tongue by the rough, brawling sprawl of English, especially American slang. Some Americans are concerned about the dying of regional accents under TV and radio's influence—they fear that soon we'll all sound as if we grew up in St. Louis.

A great language controversy in the United States is over bilingualism: Should we teach school in all the languages students come in? Or should English be the required language, and students learn it so they can learn everything else? Well, why not teach every American student at least a second language? That would be the sophisticated, cosmopolitan, useful thing to do—the geographer's solution.

Should we try to discourage the misuse of "hopefully" and the birth of such horrid words as "prioritize"? Or leave these questions open and let language evolve—as it will, whether we or the French like it or not?

1. Match the following well-known, if somewhat stale, American utterances with their utterers:

"Th-th-that's all, folks!"	Steve Martin
"*Hasta la vista, baby!*"	Bart Simpson
"Eat my shorts."	Porky Pig
"Go ahead; make my day."	Arnold Schwarzenegger character in *Terminator 2*
"What we've got here is a failure to communicate."	Strother Martin character in *Cool Hand Luke*
"Excuuuse me!"	Clint Eastwood character in *Dirty Harry*

2. What proportion of U.S. residents speak a language other than English at home?

About 1 in 450 _____

About 1 in 380 _____

About 1 in 275 _____

About 1 in 40 _____

About 1 in 7 _____

3. Which state has the most residents who speak a language other than English at home?

New York _____

Texas _____

California _____

Florida _____

Illinois _____

New Jersey _____

4. Who are the 335,000 U.S. residents—not immigrants—who speak a non-English language at home?

5. What is the most popular non-English language spoken in Louisiana, Maine, New Hampshire, and Vermont?

6. What is the most popular non-English language spoken in Montana, Minnesota, North Dakota, and South Dakota?

7. What is the most popular non-English language spoken in Rhode Island?

8. What is the most popular non-English language spoken in Hawaii?

9. What is the most popular non-English language spoken in Alaska?

10. Where did these familiar American English words originate: palooka, jazz, okay?

11. What is the most common family name in the phone book of San Jose, California?

Jones _____

Smith _____

Sanchez _____

Nguyen _____

Rodriguez _____

Gomez _____

Johnson _____

12. What is the most common language, after English, used in American newspapers?

13. Why is English considered one of the most difficult languages in the world?

14. What is "the blab of the pave"?

15. Who is Moishe Pipik?

16. What do these phrases have in common:

"Hands in the air!"

"Get lost!"

"Watch out!"

"I mean it!"

"Duck!"

"Freeze!"

17. Where is "Mei Guo"?

18. Kids aged two through fourteen have a certain linguistic good fortune. What is it?

19. What does "frontier" mean?

20. Why do so many Americans lately end their sentences on a rising intonation, as if they are ask-ing a question? Or seeking approval? Especially young Americans? Do you know what we mean?

21. What does "bayou" mean?

22. Molten rock doesn't have to form a volcano. Sometimes it just flows out in a sheet and then hardens, covering the other, older rock beneath. As the surrounding countryside erodes over the years, this hard, flat sheet is left sitting there atop a kind of, well, table. Such tables are scattered across the West. What do we call them?

23. According to the *American Heritage Word Frequency Book,* twelve of the following are the most commonly used words in written English. Knock out the one that doesn't belong:

the	you
of	that
and	it
a	he
to	for
in	I
is	

24. Which American, according to the *Guinness Book of World Records,* wrote the longest sentence ever published in English?

25. What is unique about the word "fun" in English?

26. Some Latin Americans call some North Americans *gringos.* It's not a nice word—you might call it the flip side of *spik,* which is said to derive from the Spanish-speaker's attempt to say, "I don't speak English." But where would the word *gringo* come from?

27. Early in the days of the telephone, in 1878 in Connecticut, operators answered the phone with a word that would now strike us as odd. (Hint: It wasn't "hello" and it wasn't "howdy.") What was it?

28. Match the American English word with its British English counterpart:

Baby carriage   Suspenders

Check stub   Braces

Water heater   Pram

Apartment   Flat

Garbage can   Gumboots

Wrench   Dustbin

Soccer   Cheque counterfoil

Raincoat   Geyser

Suspenders   Mackintosh

Rubber boots   Sweet

Roast   Joint

Dessert   Spanner

Garters   Football

29. Where were Pennsylvania Dutch from?

30. How did the slang word "peachy," meaning excellent, originate?

31. What kind of bug did Grace Hopper find in a circuit of her experimental computer at Harvard in 1945 that had caused it to malfunction?

32. Men have a larger vocabulary of _____ words than women; fifty-eight of them to women's mere twenty-nine, on the average. Fill in the blank.

33. What is a cheechako?

Language

1. The correct answers are as follows: Porky Pig, Schwarzenegger, Bart Simpson, Clint Eastwood, Strother Martin, Steve Martin.

2. About one in seven U.S. residents speak a language other than English at home, up from 11 percent in 1980. The top language is Spanish, with 17,339,000 speakers, followed by French with 1,702,000, German with 1,547,000, Italian with 1,308,000, Chinese with 1,249,000. The fastest-growing language in the United States is Mon-Khmer, a language of Cambodia, followed by French-Creole. The fastest-declining language is Yiddish, followed by Italian.

3. California has the most, followed in order by Texas, New York, Florida, Illinois, New Jersey, Massachusetts, Pennsylvania, Arizona, Michigan, and the rest.

4. American Indians. There are nearly 2 million American Indians in the country; 335,000 speak their own language at home, nearly half of them Navajo.

5. French.

6. German.

7. Portuguese.

8. Japanese.

9. Yupik, an Eskimo-Aleut language.

10. These words come from Africa, from the Bantu language family.

11. Nguyen, pronounced "nwin," is the most common name in San Jose. It is also the most common name in Vietnam, its country of origin. The Nguyens of San Jose came to outnumber the Johnsons in the late 1980s.

12. Spanish is the second most common language in U.S. newspapers, followed by Chinese.

13. English is hard for many reasons, but the most important may be the relation between spelling and pronunciation. In most languages, words are spelled as they sound. In English, 80 percent are not, according to Edward Rondthaler, president of the American Literacy Council. Comparing the United States with Cuba, he points out one result. "Spanish-speaking Cuba, where words are spelled as they sound, reduced adult illiteracy from 80 percent to about 20 percent in the first years of Fidel Castro's push for education. Meanwhile, adult functional illiteracy in English-speaking countries remains undiminished." Actually, in the United States it has grown.

14. The blab of the pave—the talk of the streets—was Walt Whitman's phrase for urban American slang. Most authorities would probably agree that Americans are the most profligate inventors of slang in the world, and the most deeply habituated users of it. Our "linguistic exuberance," as H. L. Mencken called it, leads us to an "excess of word-making energy."

15. Moses Bellybutton, in Yiddish; see discussions by Leo Rosten, who spells it "pupik" in *The Joys of Yiddish,* and Philip Roth, especially in *Operation Shylock.*

16. These are some of the American phrases which were to appear in a small guidebook for Japanese

visitors to the United States in the hope that more of them won't be killed by trigger-happy Americans.

17. This is America, the "beautiful country," as it is known in Chinese. Hundreds of illegal Chinese immigrants have come to U.S. shores in search of economic opportunity.

18. Between those ages, it is relatively easy for young people to learn foreign languages fluently and accent-free. After that, learning is much harder.

19. According to the U.S. Census Department, a frontier is an area with fewer than two persons per square mile living in it. In the lower forty-eight states, 13 percent of the land fits this definition. Most such areas are that way because they lack rain, and people who live there have to be tough, often holding two or more jobs just to get by. It makes for unusual versatility. In one such place, according to Dayton Duncan, author of *Miles from Nowhere,* the Lutheran minister doubled as the Presbyterian minister.

20. Well, nobody knows, but it may be a spreading form of Valley-speak—the distinctive, frequently parodied tongue of San Fernando Valley girls. Or, it may be, as linguist Deborah Tannen thinks, simply a "desire to take the other person into account." Or a request for approval? Or a demand for a response? It can also be coercive—a way of pressuring another person into agreement he or she may not feel? You know?

21. Bayou, in Choctaw, means "a sluggish stream." The bayous of southern Louisiana are so sluggish, "they seem to ooze out of the swamps," as one observer put it. "Looping and twisting, narrowing to ditches and widening into little lakes and pools, they branch and rebranch, until finally they lead to the gulf, or sometimes, back to the Mississippi itself." Bayous are a habitat of mud, favored by many plants, fish, and crustaceans that gave rise to a fabulous, French-inspired native cuisine, the Cajun-Creole.

22. Mesas—one is a mesa, Spanish for table.

23. "I" doesn't belong.

24. According to the *Guinness Book,* 1987 paperback edition, the author of the longest sentence was Jack McClintock, an obscure free-lance writer living in Miami, Florida. The sentence contained 2,403 words, "not counting numbers and counting hyphenated words as only one."

25. What's unique about "fun" in English is that it is unique *to* English—no other modern language, says the Dutch historian J. H. Huizinga, has a synonym.

26. American troops invading Mexico during the Mexican-American War of 1848 used to sing a ballad called "Green Grow the Lilacs." To Mexican ears, the first two words sounded like "gringo." The word is pejorative because, not surprisingly, the Mexicans were unhappy about the invasion.

27. Telephone operators said "Ahoy."

28. An American baby carriage is an English perambulator (or pram), a check stub is a cheque counterfoil, a water heater is a geyser, an apartment is a flat, a garbage can is a dustbin, a wrench is a spanner, soccer is football, a raincoat is a mackintosh, suspenders are braces, rubber boots are gumboots, a roast is a joint, a dessert is a sweet, garters are suspenders. And a hardware dealer is an ironmonger, a druggist is a chemist, and a bouncer is a chucker-out.

29. The so-called Pennsylvania Dutch came from Germany, not Holland; the word is a corruption of Deutsch, meaning German.

30. Early American settlers made apple cider. They also made pear cider and peach cider, and the latter was called peachy. It must have been good.

31. It was a moth, which she removed with tweezers, restoring the computer to health as if miraculously. From then on, a malfunctioning machine of any kind was said to have a "bug" in it. Until, that is, we began hearing about glitches.

32. Curse or swear.

33. *Cheechako* is the Alaskan word for tenderfoot or greenhorn.

Health

"When you don't have any money, the problem is food," wrote J. P. Donleavy, the Irish-American novelist. "When you have money, it's sex. When you have both, it's health."

Health and sickness is an interaction of humans with one another and with their environment, including other organisms we would call pests—or if they were plants, weeds.

Are some places less healthful than others? People used to think so, and it sometimes turned out they were right. It wasn't the "swamp gas" that got you, though; it was something invisible called malaria, carried by something irritating but ubiquitous and taken-for-granted, the mosquito. Early researchers suspected AIDS was a mutated version of a virus afflicting green monkeys in Uganda. How it came to afflict mothers in Minneapolis and men in Belle Glade is a complicated, tragic, fascinating, and still incomplete story, and one that intimately involved geography.

Health is like news. We tend not to pay attention to it unless it's bad.

1. What percentage of Americans who join health clubs stop attending within thirty days?

2. About 50,000 Americans are bitten by snakes each year. How many die of snakebite?

    About 6 _____

    About 150 _____

    About 300 _____

    About 3,000 _____

3. What percentage of Americans are "in a good mood" every day?

4. How many Americans die each year from insect bites?

    About 6 _____

    About 40 _____

    About 400 _____

    About 600 _____

5. How do men and women differ in the way they respond to depression?

6. In Japan, 97 percent of the dead are cremated. What percentage of Americans are cremated after death?

    5.8 percent _____

    18.5 percent _____

    27 percent _____

    41 percent _____

    56 percent _____

7. George Bush's notion of "a thousand points of light" may have sounded dim to some people, but there's no arguing that America is a nation of volunteers. We work for each other in droves, giving not only money but time to the elderly, the young, the abused, the ill, the illiterate. Many studies show that those who help others are healthier. More of us want to but are unsure how we can be helpful or what we have to give, and some public service agencies admit they are not certain what their own volunteer needs are. In many cities now, there are workshops, run by volunteers, which help potential volunteers learn how and where to volunteer. How many Americans currently do volunteer work?

    1 million _____

10 million _____

100 million _____

8. Many teenagers talk about suicide, and some of them commit it. How does this differ from the behavior of suicidal older Americans?

9. Two million older Americans are victims of _____ _____ . Fill in the blanks.

10. Seven Americans die every day waiting for an _____ _____ . Fill in the correct two words.

11. Two-part addiction question:

(a) What proportion of those who use drugs become addicted? _____

(b) What proportion of those who drink alcohol become addicted? _____

12. Who is more likely to survive a heart attack, a man who lives alone or a man who is married?

13. If you are a sedentary male and take up exercise at the age of fifty, how much longer can you expect to live as a result?

14. If all forms of cancer were eliminated overnight in America, how much longer would the average American live?

15. In infant mortality, the United States ranks _____th among nineteen industrialized countries. Fill in the figure.

16. Enforcing a ban on smoking in the workplace makes it easier for smokers to quit, and less likely that casual smokers will become heavy smokers. True or false?

17. Who tend to be more successful at quitting smoking, blacks or whites?

18. Are you more likely to die in a big city hospital with hundreds of beds, or in a rural or suburban hospital with fewer than a hundred beds?

19. In 1983, about 58 percent of Americans were overweight, but a decade later, this had changed. What proportion were overweight in 1992?

29 percent _____

38 percent _____

46 percent _____

52 percent _____

66 percent _____

20. There were 30,000 new cases of oral cancer reported in 1992. Three fourths of them were thought to have a single cause. What is it?

21. Fewer Americans are smoking. In 1955, 57 percent of males smoked; now only 28 percent do. Twenty-eight percent of women smoked back then and 23 percent still do. This may explain why _____ cancer kills more American women than _____ cancer. Fill in the blanks.

22. We know drugs can kill. But different drugs kill differently and in different numbers. Match the drug with the number of American lives it takes in a year:

Heroin	400,000
Cocaine/crack	124,000
Alcohol/car	2,000
Nicotine	2,500

23. In the former Soviet Union, one fifth of average family income is spent on alcohol. In America, the proportion is:

One sixth _____

One tenth _____

One fifteenth _____

One twentieth _____

One twenty-fifth _____

24. During Prohibition, the death rate from one particular disease fell by half. Name the disease.

25. In 1991, a study reported in the *New England Journal of Medicine* that left-handed persons die younger. What would account for such a thing?

26. What type of therapy, once considered effective, then virtually banned as medievally cruel, is making a comeback in America? (Hint: Think of Big Nurse.)

27. How many leeches do American hospitals buy per year for medicinal purposes?

28. In Bangladesh, 55 percent of the men live beyond age sixty-five. What proportion of men in Harlem do?

29. About 1,500 Americans die yearly from inhaling toxic industrial emissions. How many die from inhaling second-hand tobacco smoke?

150 _____

1,500 _____

3,000 _____

6,000 _____

14,000 _____

30. The U.S. economy is deprived of more than $16 billion a year because of something one in three Americans can't get enough of. What is it?

31. What behavioral health risk peaked in 1980?

32. What proportion of Americans are infected with a sexually transmitted viral disease (Ex: Herpes simplex or hepatitis B)?

About 1 in 1,500 _____

About 1 in 500 _____

About 1 in 50 _____

About 1 in 5 _____

33. Children are better off—healthier and better-educated—in which of these two sets of states, Column A or Column B?

A	B
New Hampshire	New York
Minnesota	Tennessee
Vermont	South Carolina
North Dakota	Florida
Iowa	New Mexico
Connecticut	Georgia
Utah	Alabama
Wisconsin	Louisiana
Nebraska	Mississippi
Maine	District of Columbia

34. Which of these cities has the highest proportion of sinus trouble sufferers?

Los Angeles _____

New York _____

Chicago _____

Philadelphia _____

Detroit _____

Boston _____

Dallas _____

Houston _____

Washington, D.C. _____

Cleveland _____

35. What proportion of Americans sometimes go outside the medical mainstream for help?

36. Half of patients, presenting a complaint, say they are interrupted by their doctor in the first:

Five seconds _____

Eighteen seconds _____

Thirty-seven seconds _____

Fifty-eight seconds _____

37. Psychologists say there are five basic dimensions of human personality: neuroticism, extroversion, openness, agreeableness, and conscientiousness. As great believers in self-improvement, Americans tend to want to grow and change, and indeed many of us do, but studies have shown that we grow within the personality we already have. That personality is mostly formed by age:

Ten _____

Sixteen _____

Twenty-one _____

Thirty _____

Forty-five _____

38. What is the relationship between television and cholesterol—beyond the fact that if you watch the news you probably worry about it?

39. What is the most pervasive flaw in the American diet?

40. Why should all newborns be screened for sickle-cell disease?

41. What proportion of older Americans suffer from diseases they would never have contracted if they'd eaten more healthful foods when they were young?

About 15 percent _____

About 45 percent _____

About 85 percent _____

42. California is a bellwether state for the United States in health. In 1984, 15.4 percent of California's population was overweight. When measured again in 1992, during the peak of the fitness craze, the proportion of California's fat folk had (slimmed to 14.7 percent; swollen to 22 percent). Pick one answer.

43. Out of 500 doctor-patient visits, how many are house calls?

44. Dr. Lester Breslow, a California public health specialist, identified "seven deadly sins"—seven habits that lead not only to early death but to chronic and costly ill health for those who manage not to die young. How many of the seven can you name? (Hint: Smoking is one.)

45. What is the most common fear (not to say phobia) among psychologists?

46. For years, one in four babies in the United States was delivered by Caesarian section. Yet the Center for Disease Control in Atlanta says 349,000 of these operations in 1991 alone were not only unnecessary, but more debilitating to the mother than natural birth. Then why are so many performed?

47. By June 1993, how many Americans had died of AIDS?

48. Does cancer kill twice as many Americans as heart disease, or does heart disease kill twice as many as cancer?

49. A study by the Harvard School of Public Health found that three quarters of adults surveyed believed "young people's safety is endangered by there being so many _____ around these days." Fill in the blank.

50. People who drink coffee have quicker reactions, remember better, and reason more clearly than those who don't. True or false?

51. How often does a sick person die in the United States while waiting for an organ for transplant?

About every four hours _____

About every for days _____

About every four weeks _____

52. The average cost of a nursing home in the United States is how much per month?

About $750 _____

About $1,575 _____

About $2,500 _____

About $3,300 _____

53. The rich are different in that they have more money. Do they also have more time?

54. What is the worst airborne killer in the United States?

55. How many living Americans have attempted suicide?

56. How much of a person's life does each cigarette smoked take away?

57. What proportion of Americans get enough exercise, according to the Centers for Disease Control and Prevention?

58. Where and in what year was the following statement made by a New Orleans surgeon, Alton Ochsner: "A distinct parallelism between the incidence of cancer of the lung and the sale of cigarettes" has been observed. "The increase is due to the increased incidence of smoking and . . . smoking is a factor because of the chronic irritation it produces."

Health

1. Seventy to 80 percent of health club joiners stop coming within a month, according to the Fitness Motivation Institute in California.

2. Of the 50,000 snakebite victims in the United States each year, only about a half-dozen die. Few American snakes are poisonous: the diamondback rattler, cottonmouth water moccasin, copperhead, and coral snake. Only about 8,000 are bitten by these, and in most cases the snake releases no venom, or releases only a little. What most often kills the snakebite victim is not venom, in any case, but shock brought on by pain and fear, says snakebite expert Maynard Cox. The introduction of a foreign protein (venom) into the body causes excruciating pain.

3. Only about 2 percent of us say we are in a good mood every day. Eleven million Americans will suffer the effects of clinical depression in a given year. One in eight will suffer it sometime during their lifetime.

4. About forty of us die from bug bites each year.

5. Women tend to deal with depression by eating. Men tend to deal with it by drinking or taking drugs.

6. Eighteen and a half percent of Americans are cremated.

7. A hundred million Americans do some form of volunteer work.

8. Americans over sixty who decide to commit suicide usually don't talk about it, they just do it, probably succeeding on the first try. The elderly have the highest suicide rate of any age group—65 percent higher than the nation as a whole and twice that of teenagers—and the numbers are no doubt vastly underreported. The people have what they believe are good reasons, and are sophisticated enough to get the means right. The main reason for this age difference is that suicide is such a taboo topic in the older generation. Many fear that talking about it will land them in a nursing home—a fate some consider literally worse than death.

9. Two million suffer from elder abuse, from their children or workers in their nursing home, according to the American Medical Association.

10. Seven die daily waiting for an organ transplant, mostly because donor organs are not available. At any given time, some 29,000 are waiting and hoping, and 25,000 die yearly for lack of a perfect match. Many times, perfectly good organs are buried with their original owner because survivors, for reasons of ignorance, superstition or the confusions of grief, refuse permission to transplant them, thus assuring two deaths instead of one. Patients who get transplanted organs and survive often find it difficult to express the depth of their gratitude. "The spark of life that's within me gives eternal life to their loved one," said one such survivor.

11. Eight percent of drug-users become addicts, while 10 percent of those who use alcohol become addicts. Fifteen million Americans use illegal drugs, and 105 million drink.

12. A married man is more likely to survive.

13. The benefits of exercise show up more clearly in quality of life than in quantity. You can expect to

live an extra ten months. You can get a lot done in that time, correct a lot of the mistakes you've made and abandon all your life's regrets (you *wish*), and you'll be stronger and happier, too. When "looking at life expectancy as it is in the Western world today, adding another year is a fairly major shift," says Dr. Graham Colditz of Harvard Medical School.

14. With the total elimination of all cancers, we would live about two years longer.

15. Nineteenth. Japan had the best rate, 5 babies per 1,000. But in America, for every 1,000 births in 1990, 9.2 babies died before age one. Black babies die at more than twice the rate of white babies, and the gap is growing. The nation's capital, whose population is largely black, has the highest infant mortality rate in the United States: 20.7 babies per 1,000.

16. True.

17. Blacks are more successful. In one study, 98 percent of blacks who quit did not resume smoking, compared with 85 percent of whites.

18. The death rate is some 25 percent higher in smaller, rural hospitals, where physicians may be less skilled and up to date.

19. Sixty-six percent were overweight in 1992, up 8 percent for the decade during which fitness consciousness rose conspicuously. Muscle weighs more than fat; could that be it?

20. Cigarettes or smokeless tobacco, according to the Surgeon General. The use of "spit tobacco" has been growing alarmingly among youngsters. One study found nearly 20 percent of high school boys had used it during the previous month.

21. Lung cancer kills more American women than breast cancer. And 46 percent of American adults still smoke tobacco.

22. Heroin kills 2,000, cocaine 2,500, alcohol and drunk driving 124,000, and nicotine 400,000.

23. One twenty-fifth of American family income is spent on alcohol.

24. Cirrhosis of the liver.

25. A faulty study and a social practice. "What you see in the population is from age five to thirty-five, the proportion that is left-handed is about 10 to 11 percent," said Dr. Marcel E. Salive, a coauthor of the newest study. "Over age sixty-five, it's more like 5 or 6 percent." But the reason for the difference, he said, is not death rates but the common practice of forcing southpaws to use their right hands as they grow. Salive's later study of 3,800 older adults found "no elevated risk of death for left-handers."

26. About 100,000 Americans get electroshock therapy each year. It is once again considered effective in specific cases. The techniques have improved since Big Nurse administered it vindictively to Jack Nicholson in *One Flew Over the Cuckoo's Nest.*

27. American hospitals purchase about 150,000 bloodsucking leeches every year.

28. Forty percent of men in Harlem live beyond age sixty-five, 15 percent fewer than in Bangladesh.

29. About 3,000 are believed to die from inhaling secondhand smoke, twice as many as are killed by any of the chemicals regulated by the Environmental Protection Agency.

30. Sleep.

31. Alcohol consumption peaked in 1980 and has been declining ever since. So too has the death rate from chronic liver disease, a largely preventable malady which dropped 23 percent in ten years but still remains the ninth leading cause of death. Alcohol is the leading cause of chronic liver disease.

32. About 1 in 5 Americans, or 56 million people, have a viral STD, according to the Alan Guttmacher Institute, and each year brings some 12 million new infections, most among persons under twenty-five. Women are the most vulnerable to sexually transmitted diseases, at least 100,000 becoming infertile each year because of them.

33. The condition of children is best in the states of the first column, A, which are ranked with the best at the top, and worst in the states of the second column, B, ranked with the worst at the bottom. The ranking comes from the *Kids Count Data Book,* compiled by the Center for the Study of Social Policy in Washington, D.C. It includes such criteria as infant mortality, percent graduating from high school, teen violent death rate, percent of children in poverty and percent of children in single-parent families. Once again, our nation's capital is at the very bottom.

34. Los Angeles has the most sinus sufferers, followed by New York and the others in decreasing proportions.

35. About a third of us spell relief with relaxation techniques, chiropractic, massage, imagery, spiritual healing, commercial weight-loss programs, macrobiotics, herbal medicine, megavitamins, self-help groups, energy healing, biofeedback, hypnosis, homeopathy, acupuncture, and folk medicine. We spend about $14 billion a year on such unconventional help. And we visit purveyors of these unorthodox treatments more often than we visit primary-care physicians. Is this because it works, or because it doesn't? Or could there be other reasons? See next question.

36. Many doctors tend to interrupt within eighteen seconds.

37. Personality is largely set by age thirty.

38. People who watch three or more hours of TV daily are twice as likely to have dangerously high levels of cholesterol (in the range of 240 mg/dl or higher) than those who watch an hour or less.

39. Too much fat. We currently get almost 40 percent of our calories from fat, which is way too high. Health experts recommend it be as low as 25 percent. Dietary fat has been connected to heart disease, some cancers, and obesity-related maladies such as diabetes.

40. Though sickle cell disease is thought to be primarily a disease of blacks, and the incidence among African-Americans is indeed very high (1 in every 375 is affected), it can and does occur in virtually all ethnic groups. In sickle cell disease, an abnormal form of hemoglobin distorts the cells into odd shapes and causes them to clog blood vessels, bringing on pain and organ damage. In America, the ethnic gene pool is so wide and deep that risk of this blood condition is very widespread.

41. The American Dietetic Association says 85 percent of Americans could have prevented diseases they now have if they'd eaten right. Listen to your mother.

42. California's fat had ballooned to 22 percent of the population by 1992.

43. One.

44. The other six habits that kill people or make them sick are: excessive drinking; being overweight; sleeping too much or too little; getting too little exercise; eating between meals; not eating breakfast. "I believe these poor health practices are indicative of a chaotic life style," Dr. Breslow said. "Regularity in one's living habits seems to be important to health and longevity."

45. The therapists' most common fear, expressed by all but one in a study of 285 of them, is that a patient will commit suicide.

46. Many, perhaps most, Caesarian sections are done more for the convenience of the obstetrician than the safety of the mother: They are quicker than waiting for a natural birth. And they are much more lucrative.

47. By June 1993, 180,000 Americans had died of AIDS—more than three times the number who died in the Vietnam War.

48. Heart disease kills nearly twice as many as cancer.

49. Guns is the answer. Public opinion has been swinging powerfully in favor of stricter gun controls. Nearly 5,000 people under nineteen were killed by firearms in 1990, according to the National Center for Health Statistics. A majority of those polled in the Harvard study, including 60 percent of the women, favored a complete federal ban on handguns, and nine out of ten favored the relatively toothless Brady Bill. In 1991, 14,265 Americans were killed by guns—two thirds of the total homicides that year.

50. A recent study published in *Psychopharmacology* found people who drink six cups a day have quicker reaction times (by 6 percent), and do 4 percent to 5 percent better on tests of reasoning and memory than those who don't drink coffee at all. Caffeine, said Dr. Martin Jarvis of the Institute of Psychiatry in London, does seem to improve concentration and vigilance.

51. An American dies every four hours waiting for an organ that did not arrive in time. Those who wish to help save lives should fill out organ-donor cards and communicate to next of kin their willingness to donate their organs.

52. The average nursing home cost in the United States was $2,500 per month in the early 1990s.

53. Indeed the rich live longer than the poor. And by 1993, when the most recent comprehensive study was conducted, the gap was widening. Americans with a family income of less than $9,000 had a mortality rate three times higher than those with a family income of $25,000 or more. This gap more than doubled between 1960 and 1986, said Dr. Gregory Pappas, an epidemiologist at the National Center for Health Statistics. "Mortality" includes death from disease, accident, occupational injuries, and violence. Some of the difference may be caused by smarter behavior among the rich, who are better-educated and who may smoke less, drink less, exercise more, stay out of the sun, and eat less fat. They also live in safer neighborhoods and don't work around dangerous machinery.

54. Soot, or particle pollution, kills 50,000 to 60,000 of us every year, more than any pollutant listed by the Environmental Protection Agency. Most soot is legal.

55. Five million have tried to take their own lives and are still alive. Chances are you may know one of them.

56. Each cigarette you smoke costs you seven minutes of life, according to government doctors.

57. About 22 percent of us are active at recommended levels. Twenty-five percent are totally sedentary.

58. Dr. Ochsner made the statement in 1945 in a speech at Duke University in North Carolina, then and now owner of a generous endowment from the sale of tobacco products.

Education

Education is an item of value. Not only do people with more education earn more money, but the more you know, the more you own the promise of an interesting life.

For geographers, education means learning more and more about our environment. We learn math so that we can measure our environment, we learn science so we can understand the world's materials and grasp its physics and chemistry. We learn history to discover the geographies of the past. And we study art and literature to learn how people portray their feelings, not only about each other, but about places. The sense of place is as critical in literature and art as it is in sports. And geography.

1. In what state do the highest percentage of children complete high school?

2. What do Columbia University and Kansas State University have in common?

3. Where do the most foreign students in America's universities come from?

4. What states have the fewest institutions of higher learning? (Two are tied for this position.)

5. About what proportion of Ivy League college students, in 1992, could name both of his or her U.S. senators?

About nine out of ten _____

About half _____

About one out of ten _____

6. What American university has the largest endowment?

7. What proportion of American adults are literate enough to function competently in our current economic environment?

About 10 percent _____

About 30 percent _____

About half _____

8. What American university has the most students on a single campus?

9. Match the service academy with its location:

U.S. Military Academy	New London, Connecticut
U.S. Naval Academy	Kings Point, New York
U.S. Air Force Academy	West Point, New York
U.S. Coast Guard Academy	Annapolis, Maryland
U.S. Merchant Marine Academy	Colorado Springs, Colorado

10. Every state but one requires youngsters to attend school until the age of sixteen, seventeen, or eighteen. Which state requires the fewest years of school attendance?

11. Where is the largest university research library?

12. What do the University of Minnesota, University of Wisconsin, University of Ohio, University of Texas, Florida State University, University of

On this map are some of the nation's leading colleges and universities marked by letters. Using the list below, link the educational institution with its letter, and give its city.

Syracuse University	L Houston
Case Western Reserve	S Evanston
University of Notre Dame	R Clemson
Duke	D Baltimore
Wellesley	C St. Louis
Vanderbilt	K Auburn
Rice	J Palo Alto
Tulane	M Syracuse
Brigham Young	W South Bend
Stanford	T Provo
Northwestern	P Cleveland
Auburn	N Williamsburg
Clemson	H Durham
Johns Hopkins	B Wellesley
William and Mary	G San Antonio
Trinity University	F Nashville
Reed College	E Portland
Washington University	A New Orleans

Answers: A Syracuse is in Syracuse; E Case Western is in Cleveland; F Notre Dame is in South Bend; G Duke is in Durham; B Wellesley is in Wellesley; H Vanderbilt is in Nashville; N Rice is in Houston; P Tulane is in New Orleans; T Brigham Young is in Provo; W Stanford is in Palo Alto; M Northwestern is in Evanston; J Auburn is in Auburn; K Clemson is in Clemson; C Johns Hopkins is in Baltimore; D William and Mary is in Williamsburg; R Trinity is in San Antonio; S Reed is in Portland; L Washington University is in St. Louis.

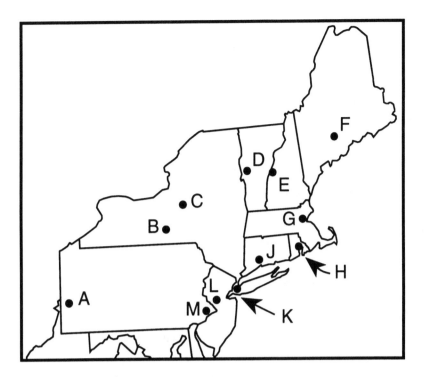

Many of the nation's oldest and most important private colleges and universities are in the Northeast. Match the educational institution from this list with its letter on the map. And give the city in which it is situated.

Dartmouth	Brown	Columbia
Harvard	Yale	Princeton
Middlebury	Colgate	Bryn Mawr
Colby	Cornell	Carnegie-Mellon

Answers: E is Dartmouth in Hanover; G is Harvard in Cambridge; D is Middlebury in Middlebury; F is Colby in Waterville; H is Brown in Providence; J is Yale in New Haven; C is Colgate in Hamilton; B is Cornell in Ithaca; K is Columbia in New York City; L is Princeton in Princeton; M is Bryn Mawr in Bryn Mawr; A is Carnegie-Mellon in Pittsburgh.

South Carolina, Michigan State University, Louisiana State University, University of Utah, and Arizona State University have in common?

13. What do these people have in common: Harry Belafonte, Cher, Mary Baker Eddy, Henry Ford, George Gershwin, D. W. Griffith, Jack London, Dean Martin, Rod McKuen, Steve McQueen, Al Pacino, Will Rogers, William Saroyan, Frank Sinatra, Orville Wright, Wilbur Wright?

14. Where in one day in America can you find about the same amount of information an average seventeenth-century Englishman or woman was likely to encounter in a lifetime?

15. What proportion of Americans over twenty-five are high school graduates?

16. What proportion of Americans are college graduates?

17. What American university scores the most federal dollars in research grants? Where is it?

18. Not long ago, four business-school professors administered a "cultural literacy" test to their students at a university whose students come from all over the nation. They were asked to identify 250 terms, such as adenoids, Gang of Four, Managua, suffrage, amortization, bear market, capital gains, lien. They got only 17.2 percent of them right. If you broke them into two groups, however, one group did slightly better than the other, the first getting all of 23 percent correct, and the second a mere 15 percent. Can you pick the groups?

Females and males _____

Liberals and conservatives _____

Whites and blacks _____

Immigrants and native-born Americans _____

19. What is a magnet school?

20. What three urban regions include the largest concentrations of colleges and universities in the country?

21. What state has the smallest number of universities?

22. Can you match these museums with their cities?

Field Museum	Tucson
Desert Museum	San Marino, California
J. Paul Getty Museum	Chicago
Huntington Library	San Francisco
Norton Simon Museum	Honolulu
Rodin Museum	Seattle
Dali Museum	St. Petersburg, Florida
Amon Carter Museum	Philadelphia
Thomas Burke Museum	Fort Worth
American Museum of Natural History	Boston
Thoreau Lyceum	New York
Museum of the American-China Trade	New Orleans
Voodoo Museum	Sarasota
Bishop Museum	Pasadena
Ringling Museum	Malibu
Palace of Fine Arts	Concord, Massachusetts

23. What country most closely links its amateur sports with its educational system?

24. While most Americans live in urban areas, we continue to believe, or seem to believe, that education takes place best in the country. Many of our leading educational institutions are in rural or pastoral settings. They were created when agricultural instruction was crucial. Here are state-supported institutions of higher learning, by region. Can you name the bucolic home of each?

NORTHEAST

University of Maine
Rutgers

University of Massachusetts
University of Connecticut
Pennsylvania State
University of Maryland
University of Vermont

SOUTHEAST

University of Virginia
University of North Carolina
University of Georgia
University of Florida
University of Mississippi

NORTH CENTRAL

University of Michigan
Indiana University
Purdue
University of Illinois
Iowa State

University of Missouri
University of South Dakota
University of Kansas

SOUTH CENTRAL

University of Arkansas
Oklahoma State University
University of Oklahoma
Texas A&M

WEST

University of Colorado
New Mexico State
Utah State
University of Wyoming
University of Idaho
Washington State
Oregon State

Answers

Education

1. Alaska, where 92 percent of those over twenty-four completed high school in 1989. Arkansas had the lowest rate: 67.6 percent.

2. Both are located in Manhattan.

3. Most—a quarter million a year in the early 1990s—came from China and Taiwan.

4. In 1988, Alaska and Nevada each had 9. New York led with 992.

5. Only about half of Ivy Leaguers can name both their senators.

6. Harvard has the largest, with a total of $4.6 billion in 1991. It costs $1.1 billion a year to run the place.

7. About half of American adults are functionally illiterate, according to a Department of Education study in 1993. Ninety million of us can't even add figures using a calculator. At the same time, an overwhelming majority of Americans said they read and write "well" or "very well." Yet it is common for employers to throw three out of four job applications directly into the trash because the applicants couldn't read or write well enough to hold down the simplest entry-level job.

8. Ohio State University in Columbus has more than 40,000 students. Think what registration must be like.

9. The Military Academy is in West Point, the Naval Academy in Annapolis, the Air Force Academy in Colorado Springs, the Coast Guard Academy in New London, and the Merchant Marine Academy in Kings Point.

10. Mississippi requires attendance only between the ages of six and fourteen.

11. In Cambridge, Massachusetts, at Harvard, which has 12 million volumes, nearly one fourth more than any other.

12. They are all situated within the metropolitan area of their state's capital.

13. They all dropped out of high school.

14. In any day's issue of *The New York Times*.

15. Seventy-seven point six percent graduated from high school.

16. Twenty-one point three graduated from college.

17. Johns Hopkins, in Baltimore, has been the nation's leader for years.

18. Liberals did a little better (23 percent) than conservatives (15 percent).

19. A magnet school is a "talent" school designated to lure students with special gifts in visual or performance arts (or, later, other special interests), and thus to motivate them better and to give parents more choice in the offspring's education. By the early 1990s, there were an estimated 5,000 magnet schools in the United States.

20. Boston, New York, and Philadelphia.

21. Wyoming, with only one, the University of Wyoming in Laramie.

22. The Field is in Chicago, the Desert Museum in Tucson, the Getty in Malibu, the Huntington in

San Marino, the Simon in Pasadena, the Rodin in Philadelphia, the Dali in St. Petersburg, the Carter in Fort Worth, the Burke in Seattle, the American Museum of Natural History in New York, the Thoreau Lyceum in Concord, the Museum of the American-China Trade in Boston, the Voodoo Museum in New Orleans, the Bishop in Honolulu, the Ringling in Sarasota, the Palace of Fine Arts in San Francisco.

23. The United States, of course. Most foreign countries have better-developed sports clubs and leagues where amateur athletes train and are developed. Some critics of American education believe the linkage between schools and sports is a main impediment to serious educational reform. Some say about half the geography instructors in American schools have the same first name: Coach.

24. NORTHEAST: Orono, New Brunswick, Amherst, Storrs, University Park, College Park, Burlington.
SOUTHEAST: Charlottesville, Chapel Hill, Athens, Gainesville, Starkville.
NORTH CENTRAL: Ann Arbor, Bloomington, West Lafayette, Champaign-Urbana, Ames, Columbia, Brookings, Lawrence.
SOUTH CENTRAL: Fayetteville, Stillwater, Norman, College Station.
WEST: Boulder, Las Cruces, Logan, Laramie, Moscow, Pullman, Corvallis.

Work

Work used to be direct. You needed something done, you did it. Then we discovered that some of us were better at some chores than at others, and we began specializing, trading my product or service for yours. Then we invented a way to carry work through time and space: money.

The basic idea of work is to create a product, but more is involved than that. Certain places are set aside for work. Work provides income, a sense of self-worth and usefulness, and creates a social context in which status is conferred. Sometimes pay and social status is commensurate with effort, skill, and the value of the product or service created; sometimes it is not (carpenters versus anchormen, for example).

Being human, and work being so central to our lives, we spend a lot of time thinking, complaining, and joking about work. That is, those of us who have jobs or portable skills do. We may be entering a time in history when jobs are no longer assured for everyone. What this may do to a nation built on something called the "work ethic" is not yet known.

1. Of which are there more in America, farmers or unemployed people?

2. What is the typical American family's income?

   About $15,000 _____

   About $25,000 _____

   About $35,000 _____

   About $45,000 _____

   About $55,000 _____

3. On what day are U.S. bosses most likely to announce layoffs?

4. Are a greater proportion of American men, or women, employed?

5. Half of the 4 million highest-paid workers in America live in the same city. Which one?

6. What state leads the United States in proportion of citizens who belong to unions?

7. During training, pro football players eat about how many calories a day?

8. How many Americans file for home-office deductions with the Internal Revenue Service?

9. Over the last twenty years, the average American worker has seen his or her working hours (increase; decrease) by the equivalent of one (day; week; month) per year. Choose the words to complete a true statement.

10. Why do we have less leisure time?

11. The second most despised professional person in America is the journalist. Who is the most despised?

12. Ten thousand people are killed and 70,000 seriously injured every year at one particular, familiar place. Where is it?

13. The 1992 work force in America was about 104 million persons. What proportion of these workers belong to unions?

    About 16 percent _____

    About 22 percent _____

    About 27 percent _____

    About 30 percent _____

14. Here are some occupations listed in a certain order. Can you determine why they've been placed in this order?

Police officer

Lawyer

Doctor

Restaurateur

Television reporter/host

Nurse

Newspaper editor/reporter

Television writer/producer

Interior decorator

Coach

Radio host/deejay

15. In real life, these are the ten most common occupations in the United States. Rank them in order of commonness.

Nurse _____

Cook _____

Salesperson _____

Janitor _____

Engineer _____

Cashier _____

Teacher _____

Secretary _____

Bookkeeper _____

Truck driver _____

16. What is the number-two cause of death at work?

17. What job competes with taxi driver for most dangerous occupation in America?

18. In Japan, about one office worker in thirty-three has his or her own office. What proportion of American office workers have their own?

Virtually all _____

About one in three _____

About one in six _____

About one in eight _____

19. If an American executive under 5 feet 5 inches earns $40,000, about how much will a six-footer earn for the same job?

About $40,000 _____

About $44,200 _____

About $48,400 _____

20. What is a telecommuter?

21. One of the following paragraphs lists occupations that are expected to gain many jobs between 1990 and 2005. The other paragraph lists some that are expected to lose many jobs in the same period. Can you tell which is which?

(a) Systems analysts, computer scientists, programmers, child-care workers, receptionists, information clerks, registered nurses, nursing aides, orderlies, restaurant cooks, lawyers, accountants, auditors, security guards.

(b) Directory assistance operators, telephone repairers, precision assemblers of electronic equipment, private household child-care workers, cable television installers, machine tool cutting operators, utilities meter readers, household help, switchboard operators, farmers.

22. Ten million to 16 million Americans have no pensions, no medical benefits, little job satisfaction, little job security, no possibility of promotion, and no future—and yet they are employed. What are they?

23. What kind of volunteer work can you do lying down?

24. What was the first U.S. multinational company?

25. Where was the equivalent of Silicon Valley in the 1830s?

26. When times are tough, people may have to take jobs for which they are overqualified—a college graduate, for instance, may take a job waiting tables or tending bar because there aren't any "better" jobs available. What proportion of Americans who graduated from college since 1980 have jobs that don't require a degree?

27. Name a good U.S. president who wasn't a lawyer.

28. Name a poor U.S. president who was a lawyer.

29. Match the American worker's average monthly earnings with the highest degree earned:

Professional	$1,672
Doctorate	4,961
Master's	1,077
Bachelor's	2,116
Associate's	492
Vocational	1,280
Some college, no degree	3,855
High school	1,672
No high school degree	2,822

30. In 1989, 17 percent of men over sixty-five were still working. What proportion were still working at that age in 1950?

23 percent _____

33 percent _____

46 percent _____

31. Public defenders are lawyers who are paid with public money to defend people charged with crimes who cannot afford an attorney. We spend $1.5 billion a year on these services. What proportion of accused persons require a public defender?

One out of ten _____

One out of five _____

Nine out of ten _____

Three out of four _____

32. How many Americans of working age are disabled?

33. What is the deadliest job in Alaska?

34. What is the American doctor's average annual income?

35. What proportion of Americans will be fired from a job at least once?

Answers

Work

1. Two percent of the American work force were farmers in the early 1990s, when this was written, and 7 percent were unemployed.

2. The average American family income, based on the 1990 census, is $35,225 a year.

3. Monday.

4. In 1993, 93.5 percent of women were employed, but only 92.6 percent of men.

5. New York City.

6. Michigan, traditional home of American auto manufacturers.

7. Pro football players eat about 4,400 calories a day during training.

8. About 4 million file for home office deductions. But there are nearly 22 million home offices nationwide, according to the National Association for the Cottage Industry.

9. The average worker now toils the equivalent of a month longer each year than twenty years ago. We enjoy less leisure time than the workers of any other industrialized country.

10. Many social critics claim the tandem effects of a materialistic culture and television advertising persuade us to work for things we don't need. We are, if they are correct, literally wasting our time.

11. The congressman, one survey found. And there must be something to it, concludes the essayist Roger Rosenblatt: "Congressmen are the only people who actively seek the company of journalists."

12. At work.

13. Just 15.8 percent of the American work force belongs to a union. In the fastest-growing segments of the work force—services and retail—less than 10 percent belong. These are also the segments with the largest concentrations of women and blacks.

14. These are the most common occupations of characters on prime-time television, with the most common first.

15. Salespersons are most common, followed by teachers, secretaries, bookkeepers, truck drivers, cashiers, janitors, nurses, cooks, and engineers.

16. Homicide is the number-two cause of death at work, right after transportation accidents and fatal contact with heavy equipment, which are lumped together. Eighteen percent of on-the-job deaths in 1992 were murders. It was worst for cab drivers, 93 percent of whose on-the-job deaths were caused by violence.

17. The clerk in a convenience store may have the most dangerous job in the country, according to the Criminal Justice Research Center. One out every hundred convenience store robberies includes a homicide, and convenience store clerks are killed at a rate two and a half times greater than police officers.

18. About one in three office workers in America have their own office.

19. The six-footer will average $4,200 more than the shorter executive. These figures apply to men, of course; it would be interesting to see how height affects the compensation of female executives (who must, of course, be shorter in order not to bump their heads on the glass ceiling).

20. A telecommuter is an employee who works at home, using computers and faxes and that sort of thing. They are the fastest-growing kind of home-workers, about 6.6 million of them so far. Even so, they represent only a sixth of the 39 million Americans who work at home, most of whom run their own businesses. Contrary to popular guessing, workers at home show higher productivity than similar workers in an office. This may be because distractions are fewer, or because of the guilty disincentives associated with goofing off on one's own time, or because they can work anytime they feel like it and they'd rather "work" than mow the lawn or clean the bathroom. There are advantages and disadvantages. Telecommuters save money on clothes, transportation, and child care, but they miss schmoozing with the gang in the office.

21. There will be fewer of the jobs in paragraph b.

22. They are the so-called nomadic service workers— those trapped in a cycle of low-paid service occupations such as cleaning, clerical work, and burger flipping. They work for minimum wage. They are the most marginal of employed persons in our society, and many have come to have little self-esteem and a lot of resentment. If we don't redesign many of these jobs in a way that creates pride and fair benefits, experts say, this worker pool will become another source of social friction.

23. Give blood.

24. Singer Sewing Machine Company was probably the first, having launched international business in the mid-1850s. By 1914 the company had more than 100,000 salespersons all around the world peddling sewing machines on the installment plan. (See Making Things.)

25. What the Silicon Valley is to the modern electronics industry, the Merrimack Valley of New England

was to the cotton textile industry, providing water power to the growing cities of Manchester and Nashua in New Hampshire, and Lowell and Lawrence in Massachusetts. Most of the rest of the power was provided by hardworking Yankee women.

26. One in five.

27. George Washington, Thomas Jefferson, Harry Truman, and John Kennedy were not lawyers.

28. You're on your own with this one. But it shouldn't be difficult.

29. Those with professional degrees earn an average of $4,961 a month; with a doctorate, $3,855; with a master's, $2,822; with a bachelor's, $2,116; with an associate's, $1,672; with some college but no degree, $1,280; with only a high school diploma, $1,077; and without a high school diploma, only $492. If you can read this, stay in school.

30. In 1950, 46 percent of men over sixty-five were still working. Social scientists have not yet explained this astonishing change in only four decades. The economy and certain social changes (corporate pensions, etc.) have a lot to do with it.

31. Three out of every four persons accused of a crime can't afford a lawyer and are defended by public-paid attorneys. Extra credit for pondering: What does this reveal about the relationship between poverty and criminality, if anything? The relationship between our acceptance of poverty and its concealed public costs? The relationship between poverty and lawyers?

32. Twenty million Americans of working age are disabled, 14 million of them are unemployed, and 9 million of those say they would like to work. To accommodate these workers, workplaces would have to be altered, but most of these alterations could be done at costs of only a few hundred dollars. Only 1 percent of such changes would cost more than $5,000.

33. If you said bush pilot, it was a good guess, but the answer is commercial fisherman. Thirty-five commercial fishermen died on the job in 1992.

34. The average doctor earns $191,000 a year—after expenses, which are considerable. But physician income varies enormously by specialty. Family doctors earn only $119,186, while cardiovascular surgeons earn $574,769 a year.

35. A third of Americans will be fired sooner or later.

Shelter

Shelter is what we use not to be outside.

Buildings often create the look of a place, thus becoming fundamental to its geography. This is especially so when they are made of native materials in traditional ways, producing such eye-pleasing, soul-stirring sights as the log cabins of the American West and Southeast or the white and blue stone buildings of the Greek Islands. In America, our adobe houses, shotgun shacks, glass-faced skyscrapers, strip malls, and garish resort hotels are closely related to our economic and architectural history, as well as to our image of ourselves. There is more to shelter than getting in out of the rain.

1. How many square feet of lawn are there in the United States for each American?

2. Who said, "Form ever follows function"?

3. What was the average price of a house in the United States in the early 1990s?

$55,000 _____

$75,000 _____

$105,000 _____

$125,000 _____

$155,000 _____

4. Name as many countries as you can that imprison a higher percentage of their population than the United States.

5. Twenty years ago, the average American house was 1,400 square feet. Today it is:

A third smaller _____

A third larger _____

About the same _____

6. To what activity do these American statistics refer?

Once a day: 9 percent

Four to six times a week: 14 percent

Two to three times a week: 28 percent

Once a week: 37 percent

Less than once a week: 10 percent

Once a month: 1.4 percent

Never: 0.06 percent

7. Where is Abraham Lincoln's tomb?

8. "Eighty percent of everything ever built in America has been built in the last _____ years." Fill in the number in this quote from James Howard Kunstler, author of *The Rise and Decline of America's Man-Made Landscape.*

9. One in four new single-family homes in the United States is a _____ _____ . Fill in the blanks.

10. Today's houses are larger than houses were in the 1950s. But one room is usually smaller than before. Which room?

11. How long did Frank Lloyd Wright labor over the design of the famous house built upon a rock over a rushing stream in Bear Run, Pennsylvania? And what was that house's name?

12. In what city was the TV sitcom *Home Improvement* set?

13. The architect Mies van der Rohe designed a famous house, too, the Farnsworth House, set in an

Illinois meadow near a woods, in 1946. What was so shockingly different about that house? (Hint: You wouldn't want to occupy it in a suburb today.)

14. Here are some rather dogmatic quotations from a modern classic book on shelter:

"A building in which ceiling heights are all the same is virtually incapable of making people comfortable."

"Everybody loves window seats, bay windows, and big windows with low sills and comfortable chairs drawn up to them."

"Balconies and porches which are less than six feet deep are hardly ever used."

Do you happen to know the book from which these were taken? (Hint: It's a fat, squat book on thin paper, with more than 1,000 pages, and the tone is unvaryingly as dogmatic as these quotations. It is also an extremely sensitive, insightful, specific and concrete summary of what makes a pleasant environment.)

15. Architecture isn't carved in stone, not even when it's a stone building. Who said, "The best friend of the architect is the pencil in the drafting room, and the sledgehammer on the job"?

16. When and where did the "balloon frame" or "stick" building method of construction originate? Why?

Barcelona, around 1655 _____

Northern Africa, before written history
began _____

London suburbs, around 1765 _____

Chicago, in the 1830s _____

17. What fixing device made balloon framing possible?

18. What was the original name of the White House?

19. What, to the early settlers of the Great Plains, was "Nebraska marble"?

20. What proportion of homeless Americans are children?

21. Where is the World Trade Center?

22. Where are these famous houses, and who lived in them?

Mount Vernon _____

Monticello _____

The White House _____

The Hermitage _____

Pickfair _____

Ashlawn _____

Biltmore House _____

Viscaya _____

Shelter

1. Five.

2. Henry Louis Sullivan, the famous turn-of-the-century Chicago architect who gave Frank Lloyd Wright his start.

3. In 1993, the average American house sold for $105,000.

4. America keeps more of its citizens in jail than any nation. This may be worth pondering. Is it something in our materialistic, individualistic culture that breeds crime? Or something in our puritanical, sanctimonious culture that breeds punishment? Are we just more efficient at catching crooks? If so, then are there still so many of them?

5. Today's new house is a third larger than a new house twenty years ago.

6. These refer to the frequency with which Americans vacuum their houses.

7. Lincoln is buried in Springfield, Illinois. What you see in Washington, D.C., is of course a memorial.

8. The answer is fifty. "And most of it," he says, "is depressing, brutal, ugly, unhealthy and spiritually degrading: the jive-plastic commuter tract home wastelands, the Potemkin village shopping plazas with their vast parking lagoons, the Lego-block hotel complexes, the 'gourmet mansardic' junk-food joints, the Orwellian office 'parks' featuring buildings sheathed in the same reflective glass as the sunglasses worn by chain-gang guards, the particle-board garden apartments rising up in every meadow and cornfield, the freeway loops around every big and little city with their clusters

of discount merchandise marts, the whole destructive, wasteful, toxic, agoraphobia-inducing spectacle that politicians proudly call 'growth.'" Kunstler doesn't think much of it.

9. Mobile home, formerly known as trailer, soon to be known, perhaps, as "manufactured home." These are the fastest-growing dwelling units in the country, though they still bear a stigma among many of us. Jimmy Buffett says they looked a lot better as beer cans. In one survey, in Virginia, they were fourth in a list of what people least wanted in their neighborhoods, after homes for the mentally or physically disabled and people on public assistance. Some 11,000 trailers were destroyed or damaged when Hurricane Andrew hit South Florida. There is criticism of their construction standards. But the average cost of a new trailer is only $30,000—about a third the cost per square foot of a site-built house. So their appeal is clear. Most mobile homes are mobile only once. Once their wheels are removed on site, 90 percent of them never move again—until a hurricane or tornado strikes.

10. Living rooms are smaller, on average, because we now spend more time in so-called "family rooms"—an entity defined mainly by the presence of a TV set.

11. Having visited the site once, Wright got hold of a topographic map, sat down with pencils, erasers, paper, tracing paper, and drew the amazingly spatially complex house in two hours. Then he scratched its name across the top: Fallingwater. The house, in Bear Run outside Pittsburgh, may be his most famous, most photographed work.

12. Detroit.

13. The Farnsworth House is a flat-roofed, rectangular box with walls of plate glass. Three years later Philip Johnson did a copycat version he called the Glass House.

14. The book is *A Pattern Language,* published in 1977 by Christopher Alexander, an American architect. Anyone thinking of designing a house could profit from its insights.

15. Frank Lloyd Wright, our great native, self-promoting, romantic genius of architecture, who also made mistakes.

16. Chicago, in the 1830s. It has largely replaced the older and handsomer, but more exacting, post-and-beam method. The stick method employed smaller, lighter, interchangeable pieces of lumber, and as can be attested by one of your coauthors, who has built two houses this way, can be easily learned by less skilled workers.

17. The mass-produced nail. Before that, houses were made of large timbers joined precisely by chiseled mortises and tenons, often fastened with wood pegs.

18. The White House was originally called the Presidential Palace. Madison called it the Executive Mansion. At first, it wasn't white; the walls were gray stone, but after a fire blackened them in 1814, the entire structure was painted white. The White House is "a copy of an Irish ducal country seat," according to Witold Rybozynski.

19. "Nebraska marble" was sod, cut into blocks and formed into houses that were dusty and gritty in summer and muddy and wet in a storm. Each lasted only a few years, but they sheltered nine out of ten Nebraska settlers in the early years.

20. One in four homeless Americans is a child.

21. The World Trade Center, two 110-story towers designed by I. M. Pei, are in New York City. They are 100 feet taller than the Empire State Building but not as tall as Chicago's Sears Tower.

22. Mount Vernon is in Mount Vernon, Virginia (George Washington slept there); Monticello is in Charlottesville, Virginia (Thomas Jefferson slept there); the White House is in Washington, D.C., the Hermitage is in Nashville, Tennessee (Andrew Jackson slept there); Pickfair is in Hollywood (Mary Pickford and Douglas Fairbanks slept there); Ashlawn is in Charlottesville (James Madison slept there); Biltmore House is in Asheville, North Carolina (the Vanderbilts slept there); Viscaya is in Miami (James Deering slept there).

Food and Drink

Food and drink can tell you a lot about local geography: What grows there, how plentiful are fuel and water, whether there is grazing land available. In America and most other developed countries, the availability of every imaginable food and drink tells us only that we are a rich nation, and that our human tendency to seek new experiences is often satisfied even in the prosaic act of feeding ourselves. Food comes readily to most of us, in such stimulating variety, no matter what the season or distance, that we've come to see food as a kind of recreation. We pursue fads, engage in pie-eating and seed-spitting contests.

But there's plenty of mystery left in food and drink. Why, for instance, do humans ''eat'' while other animals ''feed''? Why is what we eat called food, while what they eat is called feed? Why do we make exceptions for dog ''food'' and cat ''food''?

1. How much food does the average American eat over his or her lifetime?

2. In 1991, Americans bought 1.5 billion pounds of what crunchy food?

3. In 1991, the average American spent $187 on what fizzy beverage?

4. What is America's favorite fruit?

5. In February 1963, Julia Child went on television for the first time and began a great career as a demystifier of good French cooking. In what city was the TV station, and what did she cook?

6. What do Mississippi and New York have in common when it comes to drink?

7. Match the drink with the number of gallons the average American puts away in a lifetime:

Milk	1,800 gallons
Beer	2,900 gallons
Soda	2,000 gallons

8. Where in the United States should a quarter pound of hamburger cost $25?

9. We should eat at least three servings of vegetables every day, they say, and maybe some of us do. In 1982, Americans consumed an average of 81 pounds of vegetables per year. Ten years later, we ate (U-pik an answer):

   76 pounds _____

   80 pounds _____

   90 pounds _____

   100 pounds _____

   166 pounds _____

10. Americans, ever more health conscious, at least on paper, have begun worrying about the effects of caffeine. That puts most of the coffee industry on the defensive, where it struggles to maintain modest growth. But one segment of the coffee trade has actually doubled in the past ten years. Which segment is that?

11. Where did barbecue come from?

12. What is the fastest-growing segment of the beer market?

13. Drinking the water from our taps, we are trusting our local, state, and federal governments to be sure it is safe. Yet the General Accounting Office found it wasn't happening. How many states' water-inspection programs failed to evaluate all elements of their water systems that the Environmental Protection Agency recommended?

Twenty-five of the fifty states _____

Thirty-five of the fifty states _____

Forty-five of the fifty states _____

14. Whitecap flint corn, originally a species of wild grass domesticated by the American Indians, is ground up to make meal for a particular kind of breadlike food adapted by the Pilgrims from an Indian dish still eaten proudly today in Rhode Island. What is it?

15. What proportion of Americans have tried caviar?

About one in ten _____

About three in ten _____

About five in ten _____

About seven in ten _____

16. About what proportion of Americans are very interested in "gourmet" food?

About one in ten _____

About three in ten _____

About five in ten _____

17. The U.S. Department of Agriculture did a study called "Foods Commonly Eaten by Individuals: Amount Per Day and Per Eating Occasion." They found some people could really pack it away. Try to guess some of the records established, bearing in mind that these folks were not trying to impress us; this was their normal rate of consumption.

Cups of coffee in a day _____

Pancakes at a sitting _____

Pounds of beef at a sitting _____

Pounds of liver at a sitting _____

Teaspoons of sugar at a sitting _____

Slices of raw onion at a sitting _____

18. Abigail Adams, later to become First Lady, referred to an important early American import as "that bainful weed." What was this?

19. Where was the Big Mac introduced?

20. The state of New Hampshire holds the U.S. record for highest annual per capita consumption of a particular drink. What is it?

21. Match the ethnic group with the food-related health problems they often suffer:

Asian-Americans	High blood pressure deaths
Puerto Ricans	Stomach cancer
Mexican-Americans	Gallbladder, larynx cancer
Native Americans	Breast cancer
Cuban-American men	Adult-onset diabetes
African-Americans	Noninsulin-dependent diabetes

22. Can you name the two most popular ethnic entrees among average, nonethnic Americans?

23. Between 1965 and the early 1990s, U.S. consumption of a particular beverage fell from 24 gallons per capita to 19.1 gallons. Can you name the beverage?

24. During the same period, consumption of another beverage more than doubled, from 17.8 gallons per capita to an astonishing 47 gallons. Can you name this beverage?

25. A "choice" sirloin steak has about 18 percent fat. How much fat does "prime" sirloin contain?

26. Potatoes, tomatoes, peppers, and eggplant all belong to the deadly nightshade family of plants, *Solanaceae*, which produce toxic chemicals called glycoalkaloids. So why don't we die when we eat them?

27. What is America's favorite vegetable?

28. About how many pounds of potato chips does the average American eat in a year?

   1 pound _____

   2 pounds _____

   3 pounds _____

   4 pounds _____

   5 pounds _____

   6 pounds _____

29. What do they call Canadian bacon in Canada?

30. Each summer, Americans drink 204,000,000 gallons of what?

31. Where is the Dr. Pepper Museum?

32. Why shouldn't we throw rice at weddings?

   It's bad for birds _____

   It's bad for people _____

   It's bad for dogs _____

   It's bad for cats _____

   It's bad for rats _____

33. Eighty percent of all the tea drunk in the United States is drunk how?

34. What is the sweet-tea circuit?

35. In some parts of the country it is known as poor-do. It's more common name is _____ .

36. What would you consider "the quintessential meal of middle America"?

37. If Chicago with its stockyards and slaughterhouses was the city of beef early in the twentieth century, what was the city of pork?

38. What people in America eat pie three times a day?

39. Some say America has never produced a truly fine native cheese. But one food writer, Jose Wilson, has a one-word answer to that. What is the word?

40. Did you know you can make jelly out of corncobs?

41. Who claimed that the human race was born "two drinks below par"?

42. Who were the first to bring beer to America?

43. What is the difference between apple juice, apple cider, hard cider, and applejack?

44. Who said "no nation is drunken where wine is cheap"?

45. What should you eat with roast possum?

46. What is, or are, grits?

47. If you're in the Deep South and your hosts say, "Take two and butter 'em while they're hot," what will you probably reach for?

48. How do you make a mint julep?

49. What is the Smart Banana Project?

50. In the first six months of 1992, 10 percent of all the new food products introduced had a particular characteristic. What is it?

51. "In Chiavari, Italy, you can eat _____ _____ in the most expensive restaurants; try to serve them in Cleveland, Ohio, and you would be prosecuted by the Food and Drug Administration." Complete this sentence from British design critic Stephen Bayley.

52. What seafood product is consumed by Americans in greater quantities than any other except canned tuna?

53. What did a 1-pound ostrich steak cost in Texas in 1993?

54. Where was the hamburger introduced to the United States?

55. What are: (a) cola drinks, (b) breakfast cereal, and (c) ground beef?

Food and Drink

1. The average American eats 50 tons of food in a lifetime. The tonnage is vastly lower in many other countries, partly because people don't have as much, partly because they don't live as long.

2. Potato chips.

3. Beer.

4. The banana is our favorite. In 1989, a fourth of the fruit we ate was bananas.

5. Julia Child first appeared on the public broadcasting station in Boston, and she cooked boeuf bourguignonne.

6. Milk is the "state beverage" of both New York and Mississippi.

7. In a lifetime, the average American drinks 2,000 gallons of milk, 2,900 gallons of soda, and 1,800 gallons of beer. Bear in mind that some of us are doing more than our share.

8. Are we ready for the $25 quarter-pounder? Wagyu is a breed of cattle from Japan now being raised experimentally in Washington state. Wagyu produce a sort of beef known in Japan as kobe beef—the meat of cattle raised on beer and massages—which is tender, high in fat, moderate in cholesterol. It is also very labor-intensive to produce. The wagyu appear to be thriving on their Washington farms without such special attention. In Japan, kobe or wagyu beef sells for $100 a pound.

9. In 1992, Americans ate an average of 100 pounds of vegetables.

10. Sales of so-called specialty coffees have doubled. "People are converting their coffee habits from the coffee found in supermarkets to the fresher, better quality coffee found in specialty stores," says a coffee retailer. Most name-brand coffees like Folger's and Maxwell House use the common robusta coffee beans, noted mainly for their consistency from cup to cup. But first-rate specialty coffees are made from arabica beans, which have a much wider array of rich, complex, and satisfying flavors. Not to mention at least a 60 percent retail markup and prices of $10 a pound and more.

11. "Barbecue," the word, originated as *barbacoa* with the Taino Indians, who spoke Arawak, from which we get other common words including "canoe," "tobacco," and "hammock." A barbacoa was the rack of greenwood sticks upon which meat was hung above coals to cook.

12. Malt liquor—which has an alcohol content of 5.6 percent to 8 percent, versus 3.5 percent for regular beer—is the fastest-growing segment of the market. Malt liquor consumption jumped from 73.6 million cases in 1989 to 82.9 million in 1992. It is marketed, often in enormous 40-ounce bottles, mainly to poor center-city youngsters in search of a quick, cheap high. "What they [the malt-liquor pushers] do is diabolical," said the Reverend Calvin O. Butts III of the Abyssinian Baptist Church in Harlem. "They know exactly what they are doing." Brewers claim they advertise in the ghetto merely to keep a market that has always existed. Right. And cigarette companies use cartoon characters to appeal to grown-ups.

13. Forty-five of the fifty states were not fully inspecting their residents' drinking water as of early 1993.

14. Johnnycakes, or jonnycakes, a traditional Rhode Island specialty. If the meal isn't made from whitecap flint corn grown and stone ground in Rhode Island, according to a Rhode Island law adopted in 1920, it can't be called jonnycake—the spelling purists insist. Otherwise, it is a plain old johnnycake—or a cornmeal pancake.

15. About three in ten Americans have tried caviar. Fewer than one in ten liked it.

16. Only 9 percent of women and 10 percent of men are very interested in gourmet food, whatever that is. Granted it's a dumb word, and pretentious. But it only means food which has been given special attention for goodness. Are 90 percent of us so indifferent to what we eat? Or are we just afraid of a word? We certainly are not afraid to eat. See next question.

17. Some of the records were: 64 cups of coffee in a day, 27 pancakes at a sitting, 3.9 pounds of beef at a sitting, 2 pounds of liver at a sitting, 88 teaspoons of sugar at a sitting, 19 slices of raw onion at a sitting.

18. Did you guess tobacco? The answer is tea, then imported from China by British merchants. The British taxation program, much resented, led to the famous "tea party" by the Sons of Liberty in Boston Harbor in 1773. The taxes on tea and other imports had been going mostly to pay for the British troops stationed in the original colonies.

19. The Big Mac—two all-beef patties, special sauce, lettuce, cheese, pickles, and onions on a sesame-seed bun—was introduced in Pittsburgh. In May 1993, the famous sandwich celebrated its twenty-fifth anniversary.

20. Beer. The folks of New Hampshire consume an average of 51 gallons a year. Live free or die, right?

21. Asian-Americans have a higher risk of stomach cancer (possibly linked to salt-cured fish and smoked foods); Puerto Ricans have a very high rate of breast cancer (possibly linked to obesity and fatty foods); Mexican-Americans have a high rate of adult-onset diabetes (possibly linked to obesity); Native Americans have the highest rate of noninsulin-dependent diabetes in the world; Cuban-American men in Dade County (Miami) have high rates of gallbladder and larynx cancer (possibly linked to obesity and too few fresh vegetables); African-Americans have high rates of breast cancer, diabetes, and especially, deaths related to high blood pressure. American blacks have the highest rate of hypertension in the world (possibly linked partly to stress). In nearly all these cases, culturally influenced behavior plays a role in ill health.

22. In 1988, tacos and pasta were the top ethnic entrees. Nachos were the most popular ethnic appetizer.

23. Milk.

24. Sugared soft drinks. We drink 540 cans per capita per year.

25. Prime sirloin has about 30 percent fat. One reason we've all gained weight since the old, old days is the increase in fat in meat. Venison, rabbit, quail, pheasant, for instance, all have only 2 percent to 4 percent fat, and you had to be lean and fit to catch some.

26. Because we don't eat very much of the most toxic part of the plants—leaves and stems. These plants are still poisonous, however. Devouring 2.4 pounds of potato skins or eating 6.5 pounds of baked potato—of the most toxic varieties—could make an adult sick. To become ill on fried green tomatoes, we'd have to eat about 150 of them. Fully ripe tomatoes contain little or no toxin.

27. The potato is our favorite vegetable.

28. We consume an average of 6 pounds of potato chips per year. Potato chips not only are soaked in fat, they contain 250 times the salt of plain potatoes. In the food business, fat and salt are marketing devices because they add the irresistible—flavor.

29. Ham.

30. Kool-Aid.

31. The Dr. Pepper Museum is in Waco, Texas, which was also home of David Koresh and the Branch Davidians.

32. It's bad for people. Rain soaks it and turns the grains as slippery as a banana peel in a W. C. Fields movie. You could get hurt. Some say the birds clear it away after the wedding, so don't worry, but others say the rice would swell up in the bird's belly until it explodes, sending feathers, beaks, and knobby little toes all over the parking lot. But that's not so. Most birds don't even like rice. Rats and mice do, though, so don't encourage them. To please birds, throw millet, which many do like, or better yet, according to Heidi Hughes, an ornithologist with the American Backyard Bird Society, safflower seeds, which are white like rice. Many urban birds love them, but most rodents don't.

33. As iced tea—115,000,000 glasses a day.

34. The sweet-tea circuit is the Deep South, where iced tea is usually served presugared. Southerners prefer their tea sweetened five to one over unsweetened. Surveys have shown that those Southerners who do prefer unsweetened tea are urban, college-educated, and never go to church. Maybe they never sit in a porch rocker, either, which is indisputably the best place to sip iced tea, sweet or not.

35. Scrapple, that Philadelphia delicacy of ground-up pork, cornmeal, and herbs and spices shaped and cooked as a loaf.

36. Steak and potatoes.

37. Cincinnati was the city of pork, and was often called Porkopolis.

38. The Pennsylvania Dutch in and around Lancaster County, Pennsylvania, commonly have pie with—

or as—every meal. And they make pies from anything: apple, pumpkin, peach, rhubarb, coconut, cherry, lemon. There are shoofly pies—both dry and wet-bottom varieties—green-tomato pie, vinegar pie, and raisin pie—also known as funeral pie. There are fried pies, and special pies for children filled simply with milk, sugar, butter and flour. And every family traditionally has its own pet names for certain pies—poor-man's pie, grandmother's pie, eat-me-quick pie, promise pie, and schlopp.

39. Liederkranz. To those of, no doubt, unsophisticated palate, there's another word: ugh. The stinky, spreadable cheese is named for the Liederkrantz Singing Society in Manhattan, where it was first sampled. (A man we know, now a Montana rancher whom we won't name, once dropped 2 ounces of overripe Liederkranz on a Harvard classmate named McGeorge Bundy from a second-story window, and scored a direct hit.)

40. We didn't either.

41. Somerset Maugham.

42. Beer came over on the *Mayflower.* Some believe the settlers came ashore at Plymouth Rock precisely because they ran out.

43. Apple juice is apple juice. Because it contains sugar and yeast, it will ferment and become cider, its sugar becoming alcohol. Then it is hard cider, with an alcohol content of about 8 percent. In plain, nonalcoholic cider, the fermentation process has been halted by heat or preservatives. To make applejack, fermentation is allowed to continue until the alcohol is about 12 percent, and then the result is distilled into apple brandy, or applejack, or old busthead.

44. Thomas Jefferson, who also said he considered wine "a necessity of life."

45. The traditional side dish for roast possum is yams. Nothing on earth finer, so they say.

46. Grits is or are corn that has been boiled in a weak lye solution, then hulled, washed, and dried, thus becoming hominy, and then ground up, cooked in hot water within an inch of its or their already marginal existence, salted dangerously, assaulted with a fat wad of butter, and yet somehow manage(s) to turn out pretty good, though not great. Corn on the cob is better.

47. The biscuits.

48. There are dozens of ways, each considered heretical by adherents of the others. Here's one that's frankly unconventional but terrific, and which both respects and spits in the face of Old South tradition. Chill a glass. Put about six fresh mint leaves in the bottom and poke at them with a stick to release their essence. Add one drop, no more, of key lime juice. If you can't get a key lime, use a drop of lemon juice—not Persian lime. Add a bit of sugar, as little as you think you can stand— this isn't a milkshake. Crush a bunch of ice and stuff the glass full with it. Fill to the top with slightly chilled sour mash whiskey (chilling slows the melting of the ice). Crush one more mint leaf and drop it on top, so you can smell it as you sip, and add a nice big mint sprig as a garnish. If you drink three of these during the Kentucky Derby, your horse will win.

49. In the Smart Banana Project, the Rain Forest Alliance works with Central American banana producers to improve environmental protection without sacrificing productivity. In recent years, many concerned North Americans have called for a banana boycott because producers have de-stroyed lowland rain forest to grow the fruit, destroyed soil and water with pesticides and fertilizers, and produced uncontrolled amounts of waste. The Smart Banana Project developed environmental guidelines to help producers be less destructive. Those who adopt the guidelines get an "Eco-O.K." seal of approval that should provide a marketing advantage among North American environmentalists.

50. Seven hundred twelve new products were low-fat or fat-free.

51. Fish brains.

52. Shrimp. Our consumption of shrimp has doubled in the last twenty years, and per capita consumption has increased by 60 percent. We are tied with Japan's much smaller population for first place as the world's foremost shrimp-devouring nation.

53. Ostrich steak cost $29.95 a pound.

54. The hamburger first showed up in the United States in New Haven, Connecticut, in 1900, when Louis Lassen ground up lean beef (7 cents a pound at the time), broiled it, and served it on toast. No mustard, catsup, poppyseed bun, or special sauce was available, but the burger did just fine without them.

55. These are the grocery items Americans spend the most on every year, in that order.

Play

You can't surf in Iowa. There are no ski resorts in Oklahoma. Tempe, Arizona, has a swimming pool with a wave machine, but no ocean or lake. You see how geography is in everything, even—perhaps especially—recreation.

We would be a different country if we were less creative in our nonworking, nonsleeping time. Enthusiastic play—the unwillingness to depart from childhood?—seems to be fundamentally American. Evidence, of course, is all over the national landscape, in the golf courses, tennis courts, bowling alleys, ball fields, pool halls, arenas, stadiums, parks, marinas, swimming pools we haunt whenever we get the chance. A surprising number of us live less for our work or families than for the weekend, when we can get in a few games—either as players, or watchers on TV.

1. What song was recorded by, among others, Frank Zappa, John Belushi, Otis Redding, Rockin' Robin' Roberts and the Wailers, Paul Revere and the Raiders, the Kingsmen, the Fat Boys, Julie London, Mongo Santamaria, and Barry White?

2. The average cost of a round of golf in Japan is $150. What is it in America?

3. On every day of the week, television-watching is the number-one preferred use of leisure time among Americans. Socializing is number two on every day but two: Wednesday and Thursday. On those two days, Americans actually enjoy another activity second only to watching the tube. Guess what it is?

4. In what part of the United States does the largest proportion of girls participate in high school sports?

5. Where did Sonny Liston knock out Floyd Patterson for the world heavyweight boxing championship in 1961?

6. Where did Muhammad Ali knock out Sonny Liston for the world heavyweight boxing championship in 1964?

7. American teenagers who drink, smoke, steal, and are sexually active also engage more than their peers in another activity that may not be entirely good for them. Can you name it?

8. What is the southernmost team in major league baseball?

9. Where did Roger Maris break Babe Ruth's record of sixty home runs in 1961?

10. What is zydeco?

11. Can you match the stadium with its metropolitan area (most are actually in suburbs):

Wrigley Field	Detroit (Pontiac)
Astrodome	Pittsburgh
Silverdome	St. Louis
Three River	Chicago
Meadowlands	San Francisco
Robert F. Kennedy	New York (actually New Jersey)
Kingdome	Houston
Riverfront	Washington, D.C.
Busch	San Diego
Jack Murphy	Seattle
Candlestick Park	Cincinnati

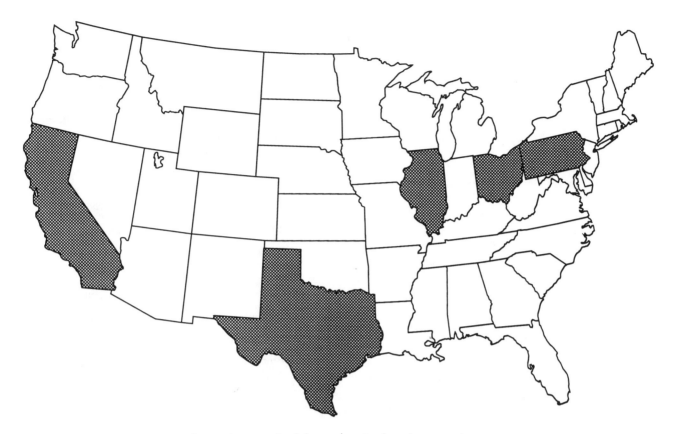

The states shaded in the map above distinguished themselves in American sports history. What did they do?

Answer: In the late 1960s, a golden age of professional football when Joe Namath, John Unitas, and Jim Brown were all playing, these were the five leading states in production of professional football players. Texas was number one, California number two, Pennsylvania number three, Ohio number four, and Illinois number five. Thanks to geographer John Rooney for this information.

12. About how many Americans own guns?

13. Measured by circulation, what is America's third-most-important magazine?

14. Why will tenors be in more demand as time goes on?

15. What is the easternmost team in the western division of American Football Conference pro football?

16. How many American homes have cable television?

17. How many hours a day of television does the average American family watch? Pick one:

1.9 hours _____

2.5 hours _____

4.4 hours _____

6.2 hours _____

7.5 hours _____

18. How many hours of TV does the average thirteen-year-old watch?

19. Who watches more television in America, blacks or whites?

20. Here are two lists of television shows from the 1992–93 season. Read the lists, then guess what distinguishes one list from the other.

(a) *Fresh Prince of Bel-Air, Roc, In Living Color, Martin, Blossom, A Different World, Out All Night, Hangin' with Mr. Cooper, Married . . . with Children.*

(b) *Roseanne, 60 Minutes, Murphy Brown, Coach, Home Improvement, Murder, She Wrote, Jackie Thomas, Monday Night Football, Love and War, Full House.*

21. A key element in every television program, especially a situation comedy, is its setting. Here are some popular programs. Match them with their hometowns. (In the real world, of course, most originate somewhere in the environs of Los Angeles, which itself, some say, does not exist.)

Murphy Brown	Boston
Designing Women	Los Angeles
Cheers	Manhattan, New York City
Full House	Cincinnati
The Simpsons	Chicago
Northern Exposure	Miami
WKRP	San Francisco
Golden Girls	Washington, D.C.
Married . . . with Children	Cicely, Alaska
Night Court	"Springfield, U.S.A."
Melrose Place	Atlanta

22. America is a mysterious place with strange criminal goings-on to keep track of from coast to coast. Luckily we have many cops and detectives to sort it all out. Match the fictional detective with his or her locale of operation.

Jessica Fletcher	Los Angeles
Father Frank Dowling	Honolulu
Sonny Crockett	San Diego
Sam Spade	New York
Harry Callahan	San Francisco
Theo Kojak	Miami
Barney Miller	Milwaukee
Simon and Simon	San Francisco
Steve McGarrett	Cabot Cove, Maine
Joe Friday	New York

23. Here is a partial list of injuries from enthusiastic participation in one particular sport in 1991. Can you guess the sport from the injuries?

Amputations: 681

Avulsions (tearing away of a body part): 344

Concussions: 236

Crushings: 399

Dental injuries: 49

Dislocations: 459

Eyeball injuries: 153

Pubic region injuries: 353

Punctures: 57

Strains and sprains: 6,893

24. Where are these post-season football bowls?

Orange Bowl _____

Rose Bowl _____

Cotton Bowl _____

Sun Bowl _____

Sugar Bowl _____

Peach Bowl _____

25. Many Americans like to climb into recreational vehicles and head out for the open road. It's a nice way to see the country, and it's cheap ($100 a day on average versus twice that for car-motel travelers).

(a) Of ten American families, how many own a recreational vehicle? _____

(b) What do 20 percent of rec-vee passengers like to do while someone else is driving? _____

26. How many comedy clubs are there in the United States?

27. What with all the fuss about healthful eating and exercise over the past dozen years, most of us are probably in terrific shape, right?

(a) For instance, only _____ out of five adults lead sedentary lives. Fill in the blank.

(b) The largest percentages of sedentary adults are found in (low-income areas; middle-income areas; high-income areas). Pick one.

(c) What percentage of Americans get no physical activity at all?

11 percent _____

15.6 percent _____

28.7 percent _____

28. A Rand Corporation study found that every mile a sedentary man walks or runs adds _____ minutes to his life. Fill in the blank.

29. Match the arena with its metropolitan area (most are actually in suburbs):

The Forum	New York
The Rosemont	San Francisco
The Omni	Sacramento
The Spectrum	Los Angeles
Madison Square Garden	Chicago
ARCO Arena	Dallas
Reunion Arena	Washington, D.C.
Capital Center	Philadelphia
The Palace	Atlanta
Cow Palace	Detroit

30. Match the ski resort with its state:

Squaw Valley	Michigan
Stowe	Utah
Park City	Vermont
Timberline	Oregon
Sun Valley	Idaho

Aspen New York

Iron Mountain California

Lake Placid Colorado

31. In most sports, the home team is thought to have a scoring advantage over the visitors. What nobody knew until it was studied was that the benefit varies with the sport. Match the percentage of home games won to the sport:

Soccer 54 percent

Basketball 57 percent

Ice hockey 69 percent

Football 61 percent

Baseball 64 percent

32. Johnny Carson was on *The Tonight Show* for thirty years. When he finally packed it in, he remarked that the earth's population had increased by (2.4 million; 2.4 billion; 24 million; 24 billion) during his tenure there. Pick one.

33. How many firearms are there in the United States?

34. In 1992, a first-run movie ticket cost an average of $6. What did one cost twenty years before?

35. There are about 200 nudist clubs in the United States. How many active nudists would you say there are?

36. The average American watches (half as much; half-again as much; twice as much; four times as much) television as the average Dane. Choose one.

37. How many Americans were in favor of gun control by 1993?

Six percent _____

Twenty-two percent _____

Forty-one percent _____

Fifty-five percent _____

Sixty-eight percent _____

38. Which kind of music do more radio stations play: top-forty pop or country music?

39. Three fourths of American television viewers would give up television for some amount of money, though many would demand millions of dollars. What's the story on the other one fourth?

40. What is the easternmost team in major league baseball?

41. In 1985, 29 percent of households with TV operated them with remote controls, sometimes called clickers or zappers. Eight years later, the proportion had grown to 87 percent. Experts believe this inevitably leads to more "channel-surfing"—zapping from channel to channel in search of something that holds one's attention—and that, in turn, led to more and more shows being produced in the fast-paced, quick-out style of MTV. Viewers' tolerance for anything that does not seem exciting was falling off, as will a reader's interest in windy, apparently pointless paragraphs such as this. But there *is* a point. Fill in the blanks in this quotation from Patricia Greenfield, a professor of psychology at the University of California at Los Angeles: "I suspect people are starting to think in _____ _____ rather than more extended arguments."

42. Where is golf's Master's Tournament held?

43. What famous baseball player was known as the Georgia Peach?

44. Where is the world's largest opera house?

45. What is the northernmost team in major league baseball?

46. One of America's great contributions to the world of thrills was introduced at the Columbian Exposition in Chicago in 1893. What was it?

47. What is the westernmost team in major league baseball?

48. If the average American child watches about three hours of television a day—as is the case—how many murders has that child witnessed by the time he or she reaches seventh grade?

49. Between 1985 and 1993, eleven American children were killed by being struck in the chest with a _____ . Fill in the blank.

50. In 1947, Jackie Robinson became the first black to play major league baseball. For what team did he play?

51. What do these places have in common:

Hapuna (Hawaii)

Bahia Honda (Florida Keys)

Kailua (Oahu)

Kapalua (Maui)

Kaunaoa (Hawaii)

Grayton Beach (Florida Panhandle)

St. George Island (Florida Panhandle)

Caladesi Island (near Tampa)

Crandon Park (Key Biscayne, Miami)

Ulua (Maui)

52. What do Ben Hogan, Rogers Hornsby, Randy Matson, Johnny Rutherford, Bubba Smith, Tris Speaker, and Babe Zaharias have in common?

53. What was the first American team to win professional hockey's Stanley Cup?

54. There were 20 million pleasure boats in the United States by 1990, three times the number in 1970. In most categories, boating deaths have declined during that period. But deaths and injuries caused by boating *collisions* have risen. Piloting a boat being at least as complex a task as driving a car, how many states require boat operators to be licensed?

55. What effect has the compact disc (CD) had on the music business?

56. What do these places have in common:

Glen Canyon National Recreation Area, Arizona/Utah

Dinosaur National Monument, Colorado

Colorado National Monument

Canyonlands National Park, Utah

Glacier National Park, Montana

Grand Teton National Park, Wyoming

Badlands National Park, South Dakota

Olympic National Park, Washington

Redwood National Park, California

Haleakala National Park, Hawaii

57. What is the favorite spectator sport in the United States?

58. Where was basketball, one of America's great recreational gifts to the world, invented?

59. What is the favorite participant sport of Americans?

60. What form of recreation do Americans spend the most money on?

61. A Filipino band playing in Anaheim, California, was doing an instrumental tune called "El Loco Cha Cha" by Rene Touzet one night in 1956 when a journeyman black musician named Richard Berry walked in and heard it. The tune had a great piano and base intro, he said, that went: duh duh duh/duh duh. Berry took the idea home and wrote a rhythm and blues calypso song about a Jamaican sailor missing his girl back in

Montego Bay. What was Berry's version of this song called?

62. Forty American hunters belong to a tiny group we might term the Ultimate Hunters, since they have had a particular, rare, and unique substance loaded into shotgun shells to be fired at animals. What is this substance?

63. Measured by bulk, in 1986, one fourth of the fiction sold in the United States was written by _____ _____ . Fill in the blanks.

64. What song was once the official song of the Leukemia Society of America, almost became the state song of Washington, and was investigated by the FBI?

Answers

Play

1. "Louie Louie" was recorded by these and many other artists. It even became elevator music, perhaps on the theory that a syrupy instrumental couldn't have the dirty lyrics the original version was thrillingly—and incorrectly—thought to have.

2. The average cost of a round of golf on an American public course is about $30.

3. Reading. Every day, we use about 30 percent of our leisure time watching television. After that, we use about 7 percent socializing. On those two days in the middle of the week, however, Wednesday and Thursday, we read—only about a fifth as much as we watch television, but we read.

4. Girls' participation in high school sports was nearing 50 percent in the mid-1980s in the North-Central region of the country: Iowa, Minnesota, Nebraska, Wyoming, North and South Dakota. It was lowest—only 10 percent—in the Southeast.

5. Chicago, Illinois.

6. Miami, Florida.

7. They watch more television.

8. The Florida Marlins.

9. Maris broke Ruth's record at Yankee Stadium.

10. Zydeco is a kind of music—a wonderfully cheery fusion of Cajun, rhythm and blues, country, and waltzes, with some jazz thrown in. You can hear a popularized version of it on Paul Simon's *Graceland* album. The real thing is even finer. The King of Zydeco, Clifton Chenier, died in 1987, but was quickly replaced by Rockin' Dopsie (pronounced DOOP-sie), real name Alton Rubin, Sr., who died in 1993 at the age of sixty-one, leaving the throne unoccupied and the crown on the shelf. Homework assignment: Listen to some good zydeco and eat some fine gumbo, regional food par excellence.

11. Wrigley Field is in Chicago, the Astrodome in Houston, Silverdome in Detroit, Three River in Pittsburgh, Meadowlands in New Jersey, Kennedy in Washington, D.C., Kingdome in Seattle, Riverfront in Cincinnati, Busch in St. Louis, Murphy in San Diego, and Candlestick Park in San Francisco.

12. About 70 million Americans own guns. Some of these are in collections, or for recreation. Nobody knows how many are not.

13. *National Geographic* is third in circulation, behind *Reader's Digest* and *TV Guide*.

14. Because as a species we are getting taller, which tends to pitch our voices lower and deeper, thus making fine tenors more and more rare.

15. The Kansas City Chiefs.

16. Sixty-one million homes had cable TV in the early 1990s.

17. The average household watches 7.5 hours of TV a day.

18. The average thirteen-year-old watches about five hours of TV a day.

19. A study conducted by Bozell Advertising found that black households watch about 50 percent

more television than white households. Blacks watched an average of 73.6 hours a week in 1992, while whites watched 50.2 hours.

20. The first list includes the ten favorite shows of blacks, most popular first. The second is the favorites of whites. Note that not one show appears on both lists.

21. *Murphy Brown* is in Washington, D.C.; *Designing Women,* Atlanta; *Cheers,* Boston; *Full House,* San Francisco; *The Simpsons,* "Springfield"; *Northern Exposure,* Cicely; *WKRP,* Cincinnati; *Golden Girls,* Miami; *Married . . . With Children,* Chicago; *Night Court,* Manhattan; *Melrose Place,* Los Angeles.

22. Fletcher is based in Cabot Cove, Dowling in Milwaukee, Crockett in Miami, Spade and Callahan in San Francisco, Kojak and Miller in New York, Simon and Simon in San Diego, McGarrett in Honolulu, and Friday is rerunning in Los Angeles.

23. People got these injuries while lifting weights.

24. The Orange Bowl is in Miami, the Rose Bowl in Pasadena, the Cotton Bowl in Dallas, the Sun Bowl in El Paso, the Sugar Bowl in New Orleans, the Peach Bowl in Atlanta.

25. (a) One of ten families own a recreational vehicle. Twenty-five million Americans vacation in them every year. (b) Twenty percent of passengers like to read.

26. In early 1990s, there were 800 comedy clubs in America.

27. (a) Three out of five adults lead sedentary lives, defined as spending less than three twenty-minute periods a week in physical activity. (b) The greatest share of sedentary adults are in low-income areas. (c) Nearly 29 percent of Americans engage in absolutely no physical activity.

28. For each mile a man walks or runs, he can live twenty-one minutes longer. Presumably, a woman would get similar benefits. However, if

it takes you twenty-five minutes to walk a mile . . .

29. The Forum is in Los Angeles, the Rosemont in Chicago, the Omni in Atlanta, the Spectrum in Philadelphia, Madison Square Garden in New York, ARCO Arena in Sacramento, Reunion Arena in Dallas, Capital Center in Washington, D.C., the Palace in Detroit, the Cow Palace in San Francisco.

30. Squaw Valley is in California, Stowe in Vermont, Park City in Utah, Timberline in Oregon, Sun Valley in Idaho, Aspen in Colorado, Iron Mountain in Michigan, Lake Placid in New York.

31. Soccer teams did best at home, winning 69 percent of the time. Basketball was next (with 64 percent), then ice hockey (61 percent), football (57 percent), and baseball (54 percent).

32. The population increased by 2.4 billion during Carson's reign, he said. He predicted that half of them would eventually have their own talk show.

33. Nobody knows. Federal officials estimate there are more than 250 million, about one for each of us.

34. A first-run movie ticket cost $1 in 1972.

35. About 40,000 Americans are committed nudists.

36. Americans watch four times as much TV as Danes.

37. Sixty-eight percent of Americans favor some form of gun control, and the percentage has been rising. But the National Rifle Association has made more noise in Congress, even without their guns.

38. Three times more stations play country than top-forty pop.

39. They wouldn't give up television for any amount of money.

40. The Boston Red Sox.

41. Sound bites.

42. At the Augusta National Golf Club in Georgia. This event was first held in 1934.

43. Ty Cobb, a Georgia native, was known as the Georgia Peach. He hit .420 for the Detroit Tigers in 1911.

44. It is the Metropolitan Opera House at Lincoln Center in New York. It seats 3,800.

45. The Seattle Mariners.

46. The Ferris wheel.

47. The Seattle Mariners.

48. The average American child has seen 8,000 murders dramatized on television by the time he or she reaches junior high school—plus more than 100,000 other acts of violence. Does this affect the child's behavior and the quality of life in our country? "Accumulated research clearly demonstrates a correlation between viewing violence and aggressive behavior," an American Psychological Association report concluded in 1992. "That is, heavy viewers behave more aggressively than light viewers. Children and adults who watch a large number of aggressive programs also tend to hold attitudes and values that favor the use of aggression to solve conflicts."

49. Baseball. These senseless deaths inspired a company, S&M Human Performance Products, to design and market a plastic heart-guard plate.

50. The Brooklyn Dodgers. The Dodgers haven't seen Brooklyn in long time.

51. These, according to *Condé Nast Traveler* magazine, are the best beaches in the United States, judged by forty criteria ranging from water quality to tropical ambience. California beaches, the judges concluded, were too crowded and polluted. "You gotta step over bodies," one said.

52. Besides being pretty good at sports, all were born in Texas.

53. First to win the Stanley Cup were the Seattle Metropolitans. It was in 1917.

54. No state, as of late 1993, required a boat driver to be licensed.

55. The CD, introduced in the United States in 1983, sold only 800,000 in that first year, while the vinyl record album sold 209.6 million, and the cassette, which had already passed the album, sold 236.8 million. Ten years later, in 1992, the picture looked and sounded decidedly different. The vinyl album barely existed, selling only 2.3 million. The cassette was still going strong, selling 366 million, but for the first time the CD had passed the cassette, selling 407 million units, and the handwriting was in the headphones, you might say. Like it or not, says one music critic, Leonard Pitts of *The Miami Herald,* we'd better come to grips with the music industry's message: "If you want to replace your scratched or broken copy of *Pet Sounds* by the Beach Boys or *That's Amore* by Dean Martin, you're going to need a CD player, because CDs are all we're selling." The CD has also, Pitts noticed, helped many people to rediscover music. "Wow," he said, referring to something he'd owned on vinyl, hadn't listened to in years, and just replayed on CD. "I'd forgotten how good that record sounds." (The album was *Cut the Cake* by the Average White Band.) What does this have to do with geography? Sound is a key part of places, environments, and regions. And sound travels, as we have seen.

56. These are the ten best places in America to get some peace and quiet, according to *Backpacker* magazine, which used criteria supplied by the Nature Sounds Society (a quiet-preservation organization) and noise measurements from the National Park Service in making its selections.

57. Football is number one when you add up high school, college, and pro.

58. James Naismith devised the game of basketball in 1891, in Springfield, Massachusetts, now home of the Basketball Hall of Fame.

59. Swimming.

60. Electronic recreation: television and radio receivers, records, and musical instruments, according to the Census Bureau. The total is five times what we spend on books and maps.

61. "Louie Louie."

62. The substance is the hunters' own cremated remains. We swear this is true.

63. Stephen King.

64. "Louie Louie."

Stuff of the Earth

Natural resources are the ones we associate with geography quizzes in school. You know, the metal ores, forest products, oil, and such. It made for a boring classroom sometimes. But it was a little like reviewing your investment portfolio—tedious and sometimes discouraging, but how else can you know how you're doing and plan your next move?

Resources include air and water and the real estate to build a business at a prime location. Time itself is a crucial resource (one we consume to zero). Resources, like investments, change location and value over time through shipping and use. And the source of resources may change. As a mine is played out or a forest laid low, we must go elsewhere for what it produced. A new resource is developed. Directions of flow may change. Political relationships may be altered as a result.

Increasingly, genetic resources, like those found in tropical rain forests, are becoming a global concern. And recycling changes resources by recovering them from products no longer useful.

1. Who are the Cornucopians and the Neo-Malthusians?

2. What are "renewable" resources?

3. "Beneath that thin layer comprising the delicate organism known as the _____ is a planet as lifeless as the moon." Complete this statement from C. Y. Jacks and R. O. Whyte by adding one crucial word.

4. Here is a line of dump trucks 3,500 miles long. They are filled with dark, rich topsoil. What do they represent?

5. "If there is magic on this planet, it is in _____ ." Complete this Loren Eiseley quotation with the name of the resource he had in mind.

6. If we could divide up all the world's water and give each person an equal amount, how big a container would you need to carry yours in?

7. We all know that 71 percent of the earth's surface is water. What proportion of that is fresh water?

8. What would happen if water froze from the bottom up instead of from the top down?

9. What human activity in the United States uses the greatest share of water?

10. A single problem wastes 20 percent to 35 percent of the public water supply in America. Can you name it?

11. Between 1970 and 1986, the amount of land in the National Wilderness Preservation System increased nearly tenfold. One state accounted for most of this. Can you name it?

12. About how many rivers are protected by the Wild and Scenic River System?

About 10 _____

About 50 _____

About 75 _____

About 100 _____

About 500 _____

13. Why bother to preserve wilderness when many of us never even go near it?

14. North America has the world's largest deposits of a very important black energy-producing material. Can you name it?

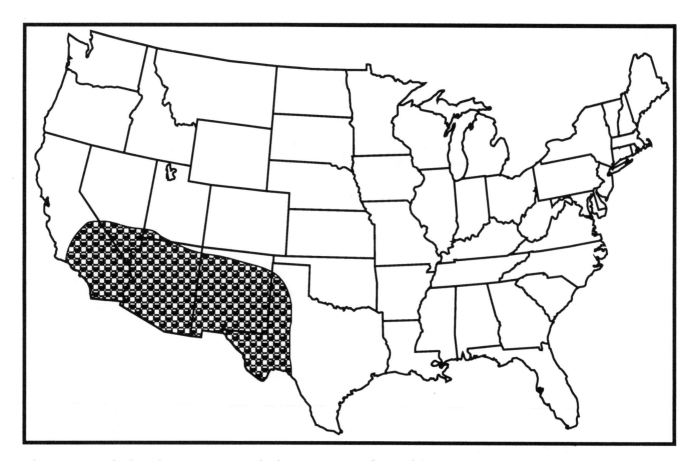

The region marked on this map contains the largest amount of one of the most important future natural resources of the United States. What is it?

Answer: Solar energy. The area marked receives intense solar radiation more than 90 percent of the daylight hours all year round.

15. What state has the most coal in the ground?

16. What is black-lung disease?

17. What was called "the most costly privately financed construction project in history"?

18. North America is lucky. Nearly every important mineral needed for modern industry is found here. But a few are not. Can you name any of them?

19. How much of the original American forest has been lost since the first colonists arrived at Jamestown in 1607?

20. What is silviculture?

21. Among natural resources, minerals are very important. They occur in two types. Can you name them?

22. Most minerals come in the form of compounds, or mixed together. But a few of them, very few, are usually found as "free elements," or standing alone. Can you name a couple of well-known and popular ones?

23. Why is it hard to find concentrated deposits of valuable or useful minerals?

24. What do magnesium, bromine, and ordinary table salt have in common?

25. Where would you find manganese nodules?

26. Bituminous coal and anthracite coal are also known respectively as _____ coal and _____ coal. Fill in the blanks.

27. What are petrochemicals?

28. Are oil and natural gas usually found together, or rather far apart?

29. Why does one state's name survive in at least two brands of oil? Name the state and the brands if you can.

30. What state produces the most natural gas?

31. Some people think that when oil reserves run low, we can get more from oil shale and tar sands. What are they talking about?

32. It is gray to pink, speckled, hard, and crystalline, and is found mainly in New England. What rock is it?

33. It is white, or varicolored, hard, either crystalline or granular, and technically a limestone. Polished smooth, it is called what?

34. There are three giant oil fields in the United States, containing more than a quarter of all the oil left in the country. Where are they?

Stuff of the Earth

1. This is a way of describing two views of the earth's resources. Cornucopians see the world as a place of unlimited resources in which conservation is unnecessary. Resources that run out *here* will be found *there,* and technology will solve our problems. If we must go to Mars for mud to make bricks, we'll go. Cornucopians have a throw-away or frontier view of the world: Use what you want now and figure out how to replace it later, trusting that the Lord, or technology, or good luck will provide. Neo-Malthusians see the world as growing ever more crowded as resources dwindle and population swells. They believe that only conservation and a sustainable-use world view will stave off ever-increasing pollution and instability. Which view do you think is nearest to correct?

2. Renewable resources are those that are not depleted if used correctly. Examples: water power, wind power, solar energy, fishing grounds, farmland. They can, however, be used incorrectly.

3. Soil (not dirt—a demeaning word for this most priceless resource). This complex mixture of minerals, rotting organic matter, water, air, and small living creatures is what creates and makes possible life on earth. Fertile soil is only 4 feet to 6 feet thick. On at least 35 percent of the world's cropland, soil is eroding faster than it is forming.

4. This much irreplaceable topsoil erodes from American croplands *every day,* resulting in plant nutrient losses of $18 billion a year. If we don't increase soil conservation, crop production may soon begin to decrease. "Civilization can survive the exhaustion of oil reserves," writes Lester R. Brown of the Worldwatch Institute, "but not the continuing wholesale loss of topsoil."

5. Water, which constitutes 70 percent of the human body, and is essential to agriculture, manufacturing, transportation, communication, and virtually every other important human activity.

6. You'd need a 77-trillion-gallon jug.

7. Only 3 percent is fresh water—and three fourths of that is locked up in glaciers or polar ice caps or so far underground that we can't use it.

8. Lakes and rivers would freeze solid when it got cold, and most aquatic life would die.

9. Agriculture uses 41 percent of water in the United States, followed by electric plant–cooling (38 percent) and industry (11 percent). The public uses only 10 percent. (In Egypt, agriculture uses 98 percent of the water.)

10. Leakage from faucets, bathtubs, showers, hoses, water mains, and pipes wastes as much as a third of our water supply.

11. Alaska, which has nearly two thirds of the U.S. wilderness area.

12. Seventy-five rivers and river systems were protected by 1987 and another 1,500 rivers and portions of rivers were under consideration.

13. The novelist Wallace Stegner has one answer. "Save a piece of country . . . and it does not matter in the slightest that only a few people every year will go into it. This is precisely its value . . . we simply need that wild country available to us, even if we never do more than drive to its edge and look in. For it can be a means of reassuring ourselves of our sanity as creatures, a part of the geography of hope."

14. If you guessed oil, nice try, but it's coal. The United States has the most, and exports the most, coal. The United States is also a major oil producer but is also the world's largest consumer of oil. And since our consumption exceeds our production, we must import a third of what we use.

15. Illinois, strangely enough, has the most coal in the ground, though it is seldom thought of as a coal-mining state. Two thirds of Illinois is underlain by coal beds.

16. Black-lung disease is a severe form of emphysema caused by breathing coal dust over a long period of time. The dust finally overwhelms the lungs' natural cleansing mechanisms. At least a quarter million American miners have spent their retirement years gasping for every breath, and have cost taxpayers $1 billion annually in disability payments because mine-safety laws were either not passed or left unenforced.

17. The Trans-Alaska Pipeline, with a final price of $8 billion, built to transport oil south from the largest oil field discovered to date in North America, from the North Slope of Alaska to the port of Valdez in the south. The 800-mile pipeline crossed three mountain ranges, thirty-four rivers, and many miles of permafrost. Nearly half of it was built 4 feet to 5 feet off the ground to prevent its heat from thawing the earth.

18. North America has no bauxite (from which aluminum is made), nor tin, nor manganese.

19. Between 45 percent and 50 percent of the original forest has been lost, but the forested area, about one third of the country, has remained roughly the same since 1920. Little of it, however, is old-growth or virgin forest.

20 Silviculture is the growing of timber on a renewable basis: tree farming.

21. Metallic minerals, such as those containing aluminum, copper, or iron, and nonmetallic minerals, such as stone, sand, and phosphates (used as fertilizer). Another way to categorize minerals is as fuel or nonfuel types.

22. Gold and silver are among the few minerals that occur as free elements.

23. Mainly because they are distributed unevenly around the earth.

24. They are the only resources abundant enough in seawater to be profitably extracted with current technology.

25. These manganese-rich rocks, about the size of yams, are found in certain spots on the deep-ocean floor.

26. Soft and hard.

27. Petrochemicals are chemicals derived from oil or natural gas. They are used in the manufacture of paints, plastics, pesticides, synthetic fibers, fertilizers, medicines, and industrial chemicals.

28. Oil and natural gas tend to be found together, the gas above the oil, beneath a dome of rock.

29. America's first oil field was near Titusville, Pennsylvania, source of the names Pennzoil and Quaker State.

30. Texas, in more ways than one, followed by Louisiana.

31. Oil shale is fine-grained rock containing hydrocarbon compounds called kerogen, which can be removed from the mined rock by heating it to high temperatures. When the vapor is condensed it forms a heavy, dark brown oil called shale oil. There are sizeable deposits of oil shale in Wyoming, Utah, and Colorado. Tar sands are a mix of clay, sand, water, and a heavy, tarlike oil called bitumen, which can be extracted by the application of heat. There are small deposits of tar sands in Utah.

32. Granite.

33. Marble, which is metamorphosed limestone. Like granite, it comes mainly from New England.

34. Prudhoe Bay, nearby Kuparuk River on the North Shore in Alaska, and the East Texas field near Houston. They also hold enormous reserves of natural gas.

Growing Things

Agriculture is one of the great topics in geography. It deals with everything from basic foods like grains and vegetables, to exotics like fancy, expensive mushrooms and seasonings, to fibers like cotton and flax and silk and wool, to narcotics like tobacco, marijuana, and cocaine, and increasingly with industrial feed stocks for the plastics and fuels industries. Agriculture also includes, of course, the husbandry of animals, which provide us, mostly without their permission, many foods and industrial products.

A tiny number of farmers produce all the food we eat plus a surplus for profitable export. High technology—in a field which is traditionally quite low-tech—is what permits this impressive feat. Great clanking machines operated by farmers now plant and pick; chemical fertilizers and pesticides do what nature can't do alone on such a scale.

Yet agriculture is more closely linked to the natural world than any other structured human activity. It is also one of the greatest consumers of land on the planet.

1. About what proportion of the earth's surface is suitable for growing crops?

2. By the 1990s, only 2 percent of Americans worked on farms. What proportion of us worked on farms in 1820?

3. How has agricultural production changed since 1820?

4. What is a "shelterbelt"?

5. In 1986, the crown prince of Liechtenstein was given $2 million of American taxpayers' money. Why?

6. Why do parts of the Great Plains appear divided into great green circles when you fly over them in summertime?

7. There are two types of agriculture practiced throughout the world. One of them is by far the most common among people in small villages of the world; the other is most common in the United States and other developed countries. Can you name or describe them?

8. In terms of energy used, expended, and gained, what is the difference between raising plant crops and raising animal crops?

9. If we divided up all the food produced in the world today, and gave each person on earth an equal share, how much would each of us get?

10. There are about 80,000 edible species of plants. Yet a mere 4 plant species comprise more of the world's total food production than all the others combined. Can you name these basic 4?

11. We use corn as a household vegetable, and make it into cereals, flour, cooking oil, syrup, and other products. But even more corn is used for something else. Can you guess what?

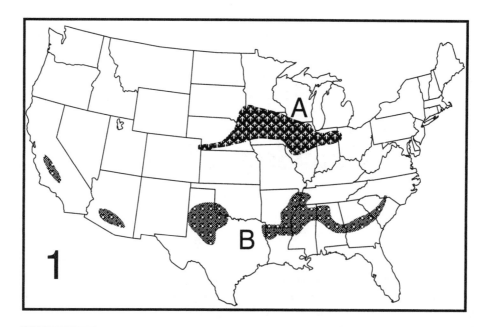

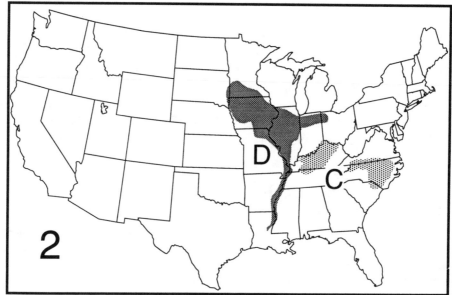

For some reason, we Americans enjoy thinking of agricultural regions as "belts." But do we know where they are? Let's see. On Maps 1 and 2, four crop regions are marked with the letters A, B, C, and D. Can you name the crop belts?

Answers: A is the Corn Belt, B is the Cotton Belt, C is the Tobacco Belt, and D is where most of the country's soybeans are grown, though we've never heard it called the Bean Belt. These "belts" merely designate regions where such crops are conspicuous, economically important, and intensely grown. Most crops—and all of these—are also grown elsewhere.

12. How many bushels of corn can an acre of land produce?

13. What was the highest price ever paid for a cow in America?

14. The world's biggest chicken farm is in Croton, Ohio, where nearly 5 million hens are laying their little hearts out. How many eggs a day would you say they produce?

15. The world's biggest mushroom farm, in Pennsylvania, produces about 23,000 tons of mushrooms a year. It is that state's largest cash crop. What does a mushroom farm look like?

16. Sixty percent of all the cranberries consumed in the United States are grown in the swampy lowlands of a single well-known peninsula. Can you name it? (Hint: It's on the east coast.)

17. What region supplies half of all the maple syrup consumed in the United States?

18. If you superimposed a map of hog cultivation in the United States onto a map of corn cultivation, you'd see they are almost coterminous. Why?

19. What are the two most important crops grown in the United States?

20. The United States is the world's largest producer of a very profitable crop that is consumed here at home and sold all over the globe. When used as intended this product is known to cause sickness and death. Yet it is not only legal and widely advertised but also subsidized by the government. Can you name it?

21. Where are the northernmost tobacco fields in the United States?

22. What single agricultural product do most Americans eat in one form or another nearly every day?

23. Besides the search for more land, what motivated the southward and westward movement of cotton plantations in the early 1800s?

24. What are Baldwin, Cortland, Northern Spy, Monroe, Rome Beauty, Winesap, McIntosh, Jonathan, and Delicious?

25. The Delicious came about by virtue of an electrifying event on Jesse Hiatt's farm near Peru, Iowa. What do you know about this?

26. What are "folk wines"?

27. Where are Vidalia onions grown?

28. On the left below is a list of the three main crops of ten selected states. Match them with their states at right.

(a) Apples, wheat, potatoes   Florida _____

(b) Tobacco, soybeans, corn   Kansas _____

(c) Corn, soybeans, hay   Washington _____

(d) Potatoes, hay, wheat   California _____

(e) Sugarcane, pineapples, coffee   Arkansas _____

(f) Peanuts, soybeans, tobacco   North Carolina _

(g) Rice, soybeans, cotton   Hawaii _____

(h) Grapes, tomatoes, cotton   Iowa _____

(i) Oranges, tomatoes, sugarcane   Idaho _____

(j) Wheat, hay, sorghum   Georgia _____

29. The farm crop creating the most value in the United States is corn. Soybeans are number three and wheat is number four. What would you guess is number two?

30. Iceberg lettuce is that common, pale, savorless, but crunchy variety of lettuce you get as "salad" in middle-American restaurants. Much of it is grown in irrigated fields near two western cities. The northern one you may know from an old movie starring James Dean, and the other you might recognize, at least by name, from TV westerns. What are these two cities?

31. What makes North Dakota wheat different from Kansas wheat?

32. What state is the biggest rice producer in the United States?

Answers

Growing Things

1. Only about 21 percent of the earth is suitable for agriculture, and it has been estimated that less than 8 percent is actually used.

2. Seventy-two percent of Americans farmed in 1820.

3. Food production doubled, and output per farmer went up eightfold. Agriculture consumes about 17 percent of all commercial energy used in the United States.

4. A shelterbelt is a row of trees planted along a field boundary to reduce soil erosion by wind.

5. The prince, not a poor man, happened to be a Texas landowner, so he got a nice chunk of federal aid intended for struggling farmers.

6. These circles are the result of center-pivot irrigation—water sprayed from a rolling spray pipe that rotates around a central well. The circles are clearly visible from airplanes and even satellite photos.

7. Subsistence agriculture, and commercial or industrialized agriculture.

8. Plant crops provide us with more food energy than it takes to raise them, thus providing an energy "profit" and becoming an energy bargain. Raising animals for food requires much more fossil-fuel energy—in feed, storage, processing, packaging, shipping, refrigerating, and cooking—than we get back from eating them.

9. Each of us would get 5 pounds of food per day—three times more than necessary to sustain life.

The root cause of starvation is not lack of food but poverty. People worldwide starve because they do not have land to grow food or money to buy it. As many as 20 million die each year because of inadequate nutrition. In some cases, Somalia being a recent one, political violence is a large contributing factor.

10. Corn, rice, wheat, and potato. And, if any of these suddenly developed a rapidly spreading disease, we'd all be in trouble.

11. More corn is fed to livestock than to people because this is the more profitable use.

12. Land—and farmers—vary in their ability to produce crops. But the world record for corn, set in 1985, was 370 bushels from an acre in Saybrook, Illinois.

13. A Holstein named Mist sold for $1.3 million in Vermont in 1985.

14. The Croton Egg Farm chickens produce about 3.7 million eggs a day.

15. Mushroom farms are dark, damp, and dank, and usually underground. The largest one in the United States, Moonlight Mushrooms, is in an old limestone mine with 110 miles of tunnels.

16. Cape Cod.

17. New England.

18. Because much of the corn crop is grown as hog feed.

19. Corn and wheat are most important, whether measured in monetary value or number of acres devoted to their cultivation.

20. Tobacco. Since the turn of the century, Kentucky and North Carolina have been the biggest tobacco producers.

21. The northernmost tobacco fields are in southwestern Wisconsin. There are others nearly as far north in the Connecticut Valley. The tobacco plant is endemic to the New World and will grow in most midlatitude and tropical environments.

22. Wheat.

23. The cotton fields moved partly to dilute the number of blacks in one place, so white slavers would feel safer.

24. These are varieties of apples grown in America.

25. Lightning struck and destroyed an apple tree on Hiatt's farm. The apple tree that grew from its roots became the basis for what evolved into the Delicious apple. Or so goes a legend of genetic alteration by electricity.

26. Folk wines, also called weed wines, are those made by people for their own consumption from whatever vegetable matter is found nearby. Perhaps the best known are elderberry wine and dandelion wine. These are much cheaper than wine made from grapes, blackberries, blueberries, gooseberries, mulberries, or raspberries, and Americans have expended no little creative thought on their invention. Weed wines are made from pansies, roses, marigolds, carnations, and daisies, from barley and rice, from mint, basil, sage, and parsley, and from potatoes, carrots, celery, beets, parsnips, spinach, tomatoes, and onions. Vidalia onions, we hope.

27. Vidalia, Georgia, where the soil helps produce onions so sweet you can eat one like a Winesap apple.

28. (a) Washington; (b) North Carolina; (c) Iowa; (d) Idaho; (e) Hawaii; (f) Georgia; (g) Arkansas; (h) California; (i) Florida; (j) Kansas.

29. Hay, which is a key commodity in the production of beef. Some students of the off-books economy believe marijuana should be high on the list as well.

30. Salinas, California, home of one of your coauthors and site of the John Steinbeck novel, and James Dean film, *East of Eden*. And Yuma, Arizona.

31. North Dakota wheat is mostly "spring wheat," while Kansas wheat is mostly "winter wheat." Spring wheat is planted in spring and matures in late summer. Winter wheat is planted in fall, lies dormant during the winter, grows in the spring, and is harvested in early summer. The winter wheat varieties are traditionally "harder," making a more flavorful bread.

32. Arkansas produces the most rice and in the early 1990s was hoping to crack the voracious and potentially profitable Japanese rice market. Japan traditionally grows nearly all its own rice and has kept its market closed to foreigners, but a crop shortage in 1993 suggested a possible opening. There was one other hurdle. Japanese diners prefer short-grained, sticky rice, while Americans like—and grow—a longer-grained, drier rice whose grains tend to separate.

Making Things

During the Depression, a belching smokestack was good news. It meant jobs. A lot of unemployed people today would not mind a reversion to the same symbolism, and we do still need manufacturing—the process of turning raw materials into finished products. Only today, the smokestack often symbolizes industrial pollution, and many of the jobs are elsewhere. But manufacturing remains a major category of work, and we all understand the central economic truth—even without Karl Marx to tell us so—that in manufacturing we are adding value to a commodity. Which makes it worth doing, and worth knowing a little about. Here are some made-in-the-U.S.A. questions about manufacturing.

1. In what year did Japanese carmakers overtake Americans as the world's number-one producer of automobiles?

   1974 _____

   1978 _____

   1980 _____

   1986 _____

   1991 _____

2. What twin cities have been known as Silicon Prairie because of their high-tech firms linked mostly with the aircraft industry?

3. What did the Hawaiian Pineapple Company change its name to in 1960 after fifty-eight years in existence?

4. What region of the United States has the most textile manufacturers?

   Pacific Coast _____

   South _____

   Midwest _____

   New England _____

   Intermountain West _____

5. What is de-industrialization?

6. What does the Rand Corporation make?

7. Few American cities have enjoyed the prosperity, or suffered the economic despair, of the auto industry as has Flint, Michigan. What brand of car made Flint famous?

8. Which American group invented a rotary harrow, a threshing machine, a circular saw, a cheese press, and a clothespin, and designed some of the simplest, most beautiful furniture in the world?

9. Who manufactured the Hula-Hoop?

10. Who invented the automobile windshield wiper, making it possible to drive safely in the rain and create drama-filled opening scenes for many American B movies as well as a few French ones, and helping Kris Kristofferson set the scene in "Me and Bobby McGee"?

11. What is a *maquiladora*?

12. Name two—or more if you can think of more—American products that employ the "disposable ammunition" school of marketing.

13. What is "consumerism"?

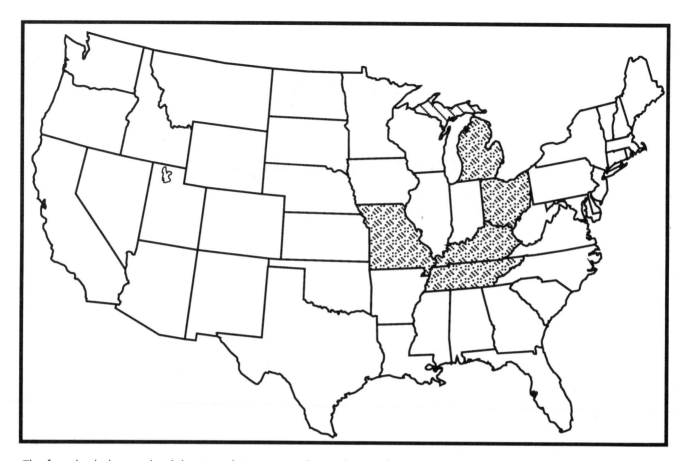

The five shaded states lead the United States in making what is often presumed to be the country's most important product. What is the product?

Answer: Automobiles. In 1993, Michigan was the number-one automaking state (with 2.9 million), Ohio was number two (with 1.7 million), Missouri was number three (with 1.2 million), Kentucky was number four (with 0.7 million) and Tennessee was number five (with 0.6 million). America's automobile manufacturing base has been moving southward as companies seek cheaper land, lower taxes, and eager labor. Much of the movement is led by Japanese multinational firms. (America's most important product, however, is educated, socially responsible children.)

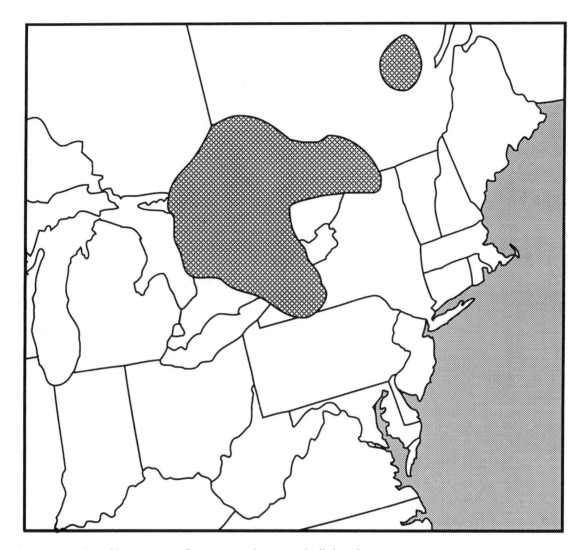

Ponder the geography of heavy manufacturing industry and all the electricity it needs. Then decide what the patterns on this map represent.

Answer: The marked areas are those regions where nitrogen dioxide deposition from acid rain is highest in the United States. It is easy to see why Canadians complain about our aerial garbage.

14. How many pigs does it take to make the 22,000 footballs consumed, as you might say, by the National Football League in a season?

15. Which metropolitan area had the greatest population decrease between 1980 and 1990? Why?

16. What did John H. Kellogg and William Keith Kellogg of Battle Creek, Michigan, and their twelve brothers and sisters have for breakfast most mornings growing up?

17. What product, still in use today and still employing the same principle, was invented in the 1850s by Linus Yale, inventor of Yale's Magic Infallible Bank Lock?

18. Where are these computer and high-tech firms headquartered?

IBM _____

Hewlett-Packard _____

Digital Equipment _____

Raytheon _____

Unisys _____

Texas Instruments _____

Apple Computer _____

Intel _____

Compaq _____

Seagate Technology _____

19. Match these oil companies with the location of their head offices:

Exxon	Bartlesville, Oklahoma
Mobil	Philadelphia
Texaco	Irving, Texas
Chevron	Houston
Amoco	Fairfax, Virginia
Shell	White Plains
Atlantic Richfield	Los Angeles
Phillips Petroleum	Chicago
Sunoco	San Francisco

20. In 1961, the brilliant industrial designer Eliot Noyes created a machine of such classic beauty and efficiency that virtually every company in America had at least one of them. The firm manufacturing this machine became one of the largest in the world. Yet today, only a few decades later, that firm has been in financial trouble for years and the product is an obsolete museum piece. Can you name the product?

21. Can you give the hometown of these well-known American product names?

Stroh's _____

Ocean Spray _____

Welch's _____

Tree Top _____

Hamm's _____

Martinelli's _____

Kraft _____

Procter & Gamble _____

Purina _____

Tabasco _____

Hershey _____

Famous Amos _____

Smucker's _____

Budweiser _____

22. What was the last vestige of true artisanship in the coachbuilding branch of the auto industry?

23. Which brand of soft drink was the first, in 1965, to use lightweight aluminum cans?

24. When Frederick Taylor, father of the time-and-motion study, was observing stokers and other workers at U.S. Steel, he learned the optimum load-weight for a man to lift in a shovel. How much weight was it?

25. Where in the United States does Honda build cars?

26. Who said, "I don't care a damn for the invention. The dimes are what I'm after!" And what invention was he talking about?

27. Where did Whitcomb L. Judson get the idea for the zipper?

28. In 1940, the U.S. Army invited 135 auto manufacturers to bid on the job of building a light utility vehicle that later became a familiar sight on American streets. Which firm got the contract?

29. How do they get the sweet pink goo inside the chocolate-covered-cherry candies?
Elves inject the goo. _____
Moonlighting watchmakers stick candy-halves together after inserting the cherry and pouring in the goo. _____
There is no goo in the candies when they are made; the goo forms as the box sits on the store shelf. _____
You got a better idea? _____

30. Where in the United States was the first Mercedes-Benz auto factory?

31. What American car, built in Dearborn, Michigan, was the first to seriously challenge the compact-car threat from abroad?

32. What was the last state to repeal World War II laws against yellow margarine (which looked a lot like the high-priced spread)?

33. What made-in-America manufactured product played the part of the blood in the shower scene in *Psycho*?

34. Where was Levi Strauss, creator of blue jeans, born?

Making Things

1. The Japanese surpassed the Americans in 1980, when they produced 11,043,000 cars. But General Motors was still the largest manufacturing company in the world.

2. Dallas-Fort Worth. Historically, Dallas was a farmer's town, while Fort Worth leaned toward ranching. The so-called Silicon Forest is around Portland, Oregon. But there is still only one Silicon Valley. In computer technical development, it leads all regions by a wide margin.

3. The Dole Corporation, after James Dole, Hawaiian Pineapple's founder.

4. The southern states have the most textile manufacturing. New England was the leader fifty years ago, but today 90 percent of the textile business is in the South, particularly North Carolina and Georgia.

5. De-industrialization is the term geographers use to describe the current economic geography of the United States, which is less and less based on manufacturing. The country now has more accountants, teachers, and lawyers than factory workers. Something similar happened years ago to farm workers—when many became factory workers.

6. Rand makes nothing. It was founded in 1943 by the U.S. Army Air Force to do research and development (RanD, get it?)—that is, to sit around and think. It was one of the first think tanks. The idea was to create a place where scientists could work on the war effort in a congenial, civilian atmosphere without a lot of pointless saluting.

7. Buick.

8. The Shakers, who also pioneered in natural-foods cookery, steaming vegetables lightly in their skins and saving the "pot likker" as stock for soups. One historian of cookery says "they typify the highest standard country-style cooking of the America that was."

9. The Wham-O Manufacturing Company of San Gabriel, California, put the 3-foot-diameter polyethylene ring on the market in 1958 as a kid's toy and adult calisthenic aid. Within six months it had forty imitators. In all, $30 million worth of Hula-Hoops were sold

10. Mary Anderson invented the wiper in 1902.

11. A *maquiladora* is one of the factories that line the Mexican side of the U.S.-Mexico border. They traditionally provided U.S.-based manufacturers with low-cost labor and limited environmental enforcement. In Ciudad Juárez, across from El Paso, there were more than 450 *maquiladoras* in 1993 as the North American Free Trade Agreement (NAFTA) was being debated.

12. The Gillette safety razor and the box camera, both of which are cheap to begin with but create a lifetime need for blades or film. The razor, invented in 1901 above a fish shop on the Boston wharves, supplanted the straight-edge razor but soon suggested its own competitor, the electric. That—an electric motor within an insulated handle—was invented by Col. Jacob Schick in 1923.

13. Consumerism is a panoply of methods by which consumer activists press manufacturers to provide safe products at a fair price. Some say it began in 1965, with publication of Ralph Nader's book *Unsafe at Any Speed*.

14. It takes 3,000 cows to make 22,000 footballs. The "pigskins" are pig-free.

15. Pittsburgh lost 162,165 people during the decade because of industrial decline.

16. The founders of the cold-breakfast-cereal industry ate hot pancakes, bacon fat, and molasses. Later John switched to something more healthful: Seven graham crackers and two apples each morning, no more, no less. Flaky? He invented corn flakes in 1894.

17. The ordinary cylinder lock most of us have on our front door.

18. IBM is in Armonk, New York; Hewlett-Packard in Palo Alto, California; Digital in Maynard, Massachusetts; Raytheon in Lexington, Massachusetts; Unisys in Blue Bell, Pennsylvania; Texas Instruments in Dallas; Apple in Cupertino, California; Intel in Santa Clara, California; Compaq in Houston; Seagate in Scott's Valley, California.

19. Exxon is in Irving; Mobil in Fairfax; Texaco in White Plains; Amoco in Chicago; Shell in Houston; Atlantic Richfield in Los Angeles; Sunoco in Philadelphia; Phillips in Bartlesville; Chevron in San Francisco.

20. The IBM Selectric I. The Selectric had a moving ball of interchangeable type that bounced mesmerizingly and efficiently. By 1975, it accounted for about 75 percent of the electric-typewriter market. Then came the personal computer and word-processing programs like WordPerfect.

21. Stroh's is a Detroit beer; Ocean Spray is in Middleboro, Massachusetts; Welch's in Concord, Massachusetts; Tree Top in Selah, Washington; Hamm's is a beer from Minneapolis; Martinelli's is in Watsonville, California; Kraft in White Plains; Procter & Gamble in Cincinnati; Purina in St. Louis; Tabasco sauce in Avery Island, Louisiana; Hershey in Hershey, Pennsylvania; Famous Amos cookies in San Francisco; Smucker's jellies and jams in Orrville, Ohio; and Bud in St. Louis.

22. Pinstriping. In the late 1980s, it was still being done by hand on some makes, usually along the beltline just under the window level to make the car look longer and lower.

23. Royal Crown, now known as RC Cola. The metal was five thousandths of an inch thick, the thickness of a magazine cover. It took Pepsi and Coke two years to catch on.

24. Twenty-one pounds—enough to move a significant load but not so much as to injure workers' backs.

25. In Marysville, Ohio.

26. Isaac Singer was speaking of the sewing machine, which he promoted into worldwide popularity by selling on the installment plan.

27. No one knows, and it's one of a few inventions that do not seem to build on others that went before. Until the arrival of the zipper, it was a world of buttons and hooks. Judson took twenty-two years to perfect it.

28. A firm called American Bantam got the contract to build the Jeep. Incredibly, only two companies had submitted bids. The other was Willys-Overland Motors, and when it became clear that Bantam couldn't keep up with production, Willys took over, improved Bantam's underpowered and unreliable vehicle, and sold the army a half million Jeeps for $740 apiece. Bantam and Willys are both out of business now. But versions of the Jeep, and knockoffs of the Jeep concept by such manufacturers as Suzuki, Toyota, Nissan, Ford, Chevrolet, and others, have become popular among trendy urbanites.

29. They coat the cherry with a paste made of sugar and a perfectly natural, harmless enzyme called invertase, which hardens. Then they dip the coated cherry into chocolate and let the chocolate firm up. They put the chocolates in those little brown-paper tutus and put them into boxes. As the boxes of candy sit around, the enzyme breaks down the sugar into sweet pink goo. Yummy.

30. Construction of the first U.S. Benz auto factory in Tuscaloosa, Alabama, began in late 1993. The firm had been building trucks here for years.

31. The Ford Falcon, in 1958, was the first serious attempt to challenge the Volkswagen. A few years earlier, Kaiser's Henry J had the right idea, but Americans—except for one of your authors' fathers—didn't buy Kaisers, Frasers, or Henry Js. (Most people today have never heard of these comically bulbous cars.) Six years later, Lee Iacocca's great coup, the Ford Mustang, made his reputation forever. The Mustang was merely a warmed-over, tarted-up, Falcon.

32. Wisconsin, a great dairy state, did not permit yellow margarine to be sold until 1967. Even then, at the behest of the dairy industry, it continued to impose special taxes on margarine.

33. Hershey's chocolate syrup, from Pennsylvania. The famous Alfred Hitchcock thriller was filmed in black and white.

34. The creator of Levi's was born in Buttenheim, Bavaria.

Trading Things

Business is mainly about trading. We tend to think of trade items as tangible, like new cars, truckloads of orange juice, shiploads of oil, chrome-trimmed turbo widgets. But increasingly, trade is in information, on many topics, packaged in many forms and vended by people like lawyers and anchorpersons. More and more, people get paid for what they know about how the world works. This, actually, is the traditional business of geographers.

The geography of business has to do with the obvious: where materials originate, where they are transformed or repackaged into "products," where the natural customer pool may be for such a product. Is it smart to be near your customers, to serve them better, or smarter to be close to your source of raw materials, and save on transportation costs? Or even smarter to set up shop in a beautiful community with nice weather, recreation opportunities, and good schools that keep your workers happy so they'll be more productive and maybe even willing to work for less? Most people would rather live near a beach in California than a swamp in Mississippi.

What if your community isn't attracting business? Well, maybe if you build them something—a stadium, a public transport system—they will come.

A clever geographer might help answer such questions.

1. What are the three largest industries in America?

2. According to a Department of Agriculture study, what determines how much a shopper spends in a supermarket?

    How far the shopper walks _____

    How long the shopper shops _____

    How low the prices are _____

    How well-advertised the store is _____

    Coupons _____

3. Where is the Western Hemisphere's largest casino? Who operates it?

4. What large, nationwide business started out with a single store in Kemmerer, Wyoming, a jump-off point on the Oregon Trail?

5. What were the three largest business firms in America in 1991?

6. Why should John Johnson be ranked as one of the country's most influential businessmen, even though you may never have heard of him?

7. What American port was the gateway to Latin America for decades until Miami took that title in 1990?

8. What are the three largest retailing companies?

9. In terms of total output (in 1991), which state produced the most:

Corn	Texas
Wheat	Iowa
Cotton	Idaho
Potatoes	Iowa
Tobacco	Kansas
Cattle	North Carolina
Pigs	Texas

10. How much does a new grocery shopping cart cost?

11. Which state received the most money from the federal government in 1991?

12. Which U.S. port is the largest, in terms of tonnage handled?

13. Rank these three North American countries for crude petroleum production:

Canada _____

Mexico _____

United States _____

14. The average value of the goods and services produced each year by a Japanese worker is $38,200. What is the value produced by an American worker?

$23,400 _____

$28,250 _____

$35,600 _____

$39,200 _____

$42,550 _____

$49,600 _____

15. What metropolitan area in the United States has the most commercial office space?

16. What proportion of American managers are incompetent, according to a psychologist who studied the subject?

17. Match the hotel with its city:

Fairmont	Chicago
Disneyland	Miami Beach
Benson	Palm Beach
Drake	San Francisco
Fountainbleau Hilton	Anaheim
Plaza	New York
Waldorf Astoria	Portland, Oregon
Mark Hopkins	Chicago
Palmer House	San Francisco
Breakers	New York

18. Tourism has long been one of the most important industries of the United States. What's new is that so many of the tourists now come from foreign countries. How much, on average, is each foreign tourist worth to the federal government?

19. Match these transport businesses with their home offices:

General Motors	Atlanta
Ford	Memphis
Chrysler	Atlanta
Boeing	Elk Grove, Illinois
United Airlines	Detroit
United Parcel Service	Dearborn, Michigan
Federal Express	Eagan, Minnesota
Delta Airlines	Arlington, Virginia
Northwest Air	Seattle
USAir	Highland Park, Michigan

20. Match these retailing businesses with their home offices:

Sears, Roebuck	Troy
Wal-Mart	Oakland
K mart	Minneapolis
Kroger	Dallas
J. C. Penney	Chicago
Dayton Hudson	Bentonville
Safeway	Cincinnati
May department stores	Oak Brook
Albertson's	Paramus
McDonald's	St. Louis
Toys "R" Us	Boise
Macy's	Seattle
Meier & Frank	Dallas
Bonmarché	Portland, Oregon
Burdine's	Chicago
Neiman-Marcus	New York
Marshall Field	Miami

21. Why haven't the baby boomers, the 72 million Americans born between 1946 and 1964, become the enormous political force their numbers suggest they could be?

22. Where was the first robbery of Jesse James and his gang, who later graduated to banks and trains?

23. What was the first U.S. multinational company?

24. The S. S. Kresge chain-store empire began as a five-and-dime which opened in what city in 1898?

25. The idea of cup-sizing for brassieres was devised by Leona Gross Lax for the Warner Brothers Company of what city?

26. What and where was the first store to be fully air-conditioned?

27. With backing from syndicate boss Meyer Lansky, Benjamin "Bugsy" Siegel built the Flamingo Hotel and resort, changing the economy of an entire state. How and where?

28. Where was the TV dinner invented?

29. What product of the earth helped Ajax beat Bon Ami? Where does Comet come in?

30. What and where was the first discount retail store?

31. Why did sales of oregano in the United States rise 5,200 percent between 1948 and 1956?

32. Many of America's wealthiest families acquired their money in a previous generation by manufacturing and merchandising. Try to match these famous family names with (a) their best-known product or business and (b) their hometown:

Andersen	Beer	Pittsburgh
Bacardi	Sporting goods	St. Louis
Bean	Oil	Bayport, Minnesota
Busch	Rum	Morristown, New Jersey
Dayton	Windows	San Juan, Miami
du Pont	Department stores	Philadelphia
Lilly	Medicines	Freeport, Maine
Mead	Chemicals	Minneapolis
Mellon	Paper	St. Louis
Mennen	Publishing	Wisconsin Rapids
Pew	Toiletries	Indianapolis
Pulitzer	Banking	Wilmington, Delaware

33. What American athlete, in the early 1990s, was earning more per year than any other athlete in history?

Trading Things

1. Automobiles, supermarkets, and tourism, in that order, produce the most income in the United States.

2. How far the shopper walks.

3. The Foxwoods High Stakes Bingo and Casino in Ledyard, Connecticut, was the largest in late 1993. The Mashantucket Pequot Indian tribe owned it. The growth of "Indian gambling" in many areas has Las Vegas and Atlantic City worried. Indian gambling operations not only don't pay federal and state taxes, they are largely beyond the reach of city and state laws.

4. J. C. Penney.

5. General Motors, Exxon, and Ford Motor Company, in that order.

6. John Johnson, of Chicago, produces *Jet, Ebony,* and *EM*—the country's leading magazines serving an African-American audience.

7. New Orleans.

8. Sears, Wal-Mart, and K mart, in that order. No other company even comes close to these three.

9. Iowa produced the most corn, Kansas the most wheat, Texas the most cotton, Idaho the most potatoes, North Carolina the most tobacco, Texas the most cattle, and Iowa the most pigs. By total value, California is the top agricultural state in the country.

10. A shopping cart costs about $100, and the loss of thousands of them has devastated supermarket profits, which average only about 1 percent. Homeless folks use the carts as their only wheels, and retirees who don't have cars often trundle their groceries home in the carts and then leave them on the street. This has created incentive for a new business: shopping-cart bounty hunters. These entrepreneurs round up the missing carts and charge the supermarket about $20 for a truckload of twenty of them.

11. California got the most federal money, which seems fair in that California paid the most taxes.

12. The Port of South Louisiana, as it's called, which includes all of the ports south of Baton Rouge, handled the most. The total is high because of foreign imports of all sorts, but particularly oil and trade with the Caribbean. And most of the cargo from Mississippi barges is loaded onto ocean vessels here.

13. The United States still produces by far the most crude oil, twice as much as Mexico and five times as much as Canada. We use a lot more, too, which is why we import so much.

14. The average American worker produced $49,600 in goods and services in 1992.

15. New York by a wide, wide margin. Washington, D.C., is second with only a fourth as much.

16. Sixty to 70 percent of American bosses are not good managers, according to Dr. Robert Hogan of the University of Tulsa. He blames the selection process in which workers are promoted to supervisor because of their skills, but then promoted to middle management because they curry favor with the boss.

17. The Fairmont and Mark Hopkins are in San Francisco, the Disneyland in Anaheim, the Benson in Portland, the Drake and Palmer House in Chicago, the Fountainbleau Hilton in Miami Beach, the Plaza and Waldorf Astoria in New York, and the Breakers in Palm Beach.

18. The federal treasury racked up an average of $89 in taxes from each visitor in 1992.

19. GM is in Detroit, Ford in Dearborn, Chrysler in Highland Park, Boeing in Seattle, UAL in Elk Grove, UPS in Atlanta, FedEx in Memphis, Delta in Atlanta, Northwest in Eagan, USAir in Arlington.

20. Sears is in Chicago; Wal-Mart in Bentonville, Arkansas; K mart in Troy, Michigan; Kroger in Cincinnati; J. C. Penney in Dallas; Dayton Hudson in Minneapolis; Safeway in Oakland; May in St. Louis; Albertson's in Boise; McDonald's in Oak Brook, Illinois; Toys "R" Us in Paramus, New Jersey; Macy's in New York; Meier & Frank in Portland, Oregon; Neiman-Marcus in Dallas; Burdine's in Miami; Bonmarché in Seattle; and Marshall Field in Chicago.

21. The answer may be that there are two sets of boomers: older boomers and younger boomers. The older ones regard the younger as "hopelessly materialistic . . . utterly lacking in commitment," writes Katherine S. Newman, author of *Declining Fortunes: The Withering of the American Dream.* And the younger ones view the older boomers, the flower-power generation, as "people whose attitudes were fine for their own time but out of step with today's realities." These groups' ideas are so far apart that they cannot see their shared economic interest as "the first generation since the Great Depression that can expect to have a *lower* standard of living than its parents."

22. The James boys' first job was robbing the receipts of a local fair in Kansas City. Most of their activities were confined to the Midwest.

23. The Singer Sewing Machine Company began its spread in the middle 1850s and by 1914 had 100,000 salespersons all around the world. Selling on the installment plan was key to its success.

24. Detroit. The Kresge chain grew to 677 stores in twenty-seven states during the first forty years of its existence. The firm's descendent is K mart.

25. Bridgeport, Connecticut. No relation to the movie company.

26. Tiffany & Co. in New York was the first to open a fully air-conditioned store, in 1940.

27. Siegel built the Flamingo Hotel and Casino in Las Vegas. Legal gambling changed Nevada forever. Bugsy, though, was shot to death by other mobsters the following year. Warren Beatty played him in the movie.

28. The TV dinner was invented in Omaha, Nebraska, by some genius at C. A. Swanson & Sons, maker of frozen meat pies.

29. Ajax hit the market in 1947 containing silica sand as its abrasive cleaning agent. Though it tended to scratch objects more than the feldspar used in earlier cleansers, consumers loved it, and soon Ajax was outselling Bon Ami and Old Dutch Cleanser. When Comet came along nine years later, it also contained silica sand—and was soon the best-selling cleanser.

30. The first discount store was E. J. Korvette in New York City. It opened in 1948 and vanished several years ago.

31. In a word: pizza. American servicemen in Italy came to love the food.

32. Andersen: windows; Bayport, Minnesota
Bacardi: rum; San Juan and Miami
Bean: sporting goods; Freeport

Busch: beer; St. Louis
Dayton: department stores; Minneapolis
du Pont: chemicals; Wilmington
Lilly: medicines; Indianapolis
Mead: paper; Wisconsin Rapids
Mellon: banking; Pittsburgh
Mennen: toiletries; Morristown, New Jersey

Pew: oil; Philadelphia
Pulitzer: publishing; St. Louis

33. Arnold Palmer. His income was millions every year, only a small portion of it from golf; he earned far more from endorsements and business ventures.

Crime and Punishment

Americans fear crime. But we also have a deep and lasting affection for it, and, building on English tradition, we have created a great literature on the subject—or rather, many literatures, from high to low. Perhaps we're so fond of crime because it is the perfect expression of our true national faith: individualism.

Crime is eminently geographical. Date-rape statistics are high in college towns. Coastal city cops bust a lot of drug smugglers. White-collar crime abounds in financial districts. Violent crime is common where people are poor and feel cheated. Crime has causes. It is largely a matter of economy, psychology, and opportunity—and at least two of these have geographical content.

We may think about crime more than we need to, because in fact most of us live in a safe society. It could be safer if we fixed visible, soluble problems such as an irrational judicial system and the proliferation of guns—53 percent of murders are carried out with easy-to-buy handguns. We could also keep bad people locked up, if we could bring ourselves to believe there actually is such a beast. On the other hand, the most repressive police states in the world are the safest to walk in at night. Do we want to pay the price? Fortunately, that decision is political, not geographical.

1. In late 1993, sixty-five Americans a day were being killed by one thing. What was it?

2. A woman is raped in America every seven minutes. One of four women will be sexually assaulted sometime in her life. What percentage of the victims know their attacker?

    18 percent _____

    29 percent _____

    58 percent _____

    78 percent _____

3. Who commits most of the violent crimes against whites in America—whites or blacks?

4. Who commits most of the violent crimes against blacks in America—blacks or whites?

5. What percentage of criminals are responsible for 80 percent of crimes in America?

    10 percent _____

    20 percent _____

    30 percent _____

    40 percent _____

6. We don't like to think about it, but lots of criminals get away with their crimes. Match the crime with the percentage of them "cleared" by arrest:

Murder	13 percent
Aggravated assault	24 percent
Rape	52 percent
Robbery	57 percent
Burglary	67 percent

7. Are most American burglaries committed in the daytime or at night?

8. What is the difference between a serial killer and a mass murderer?

9. Which category of crime has not increased in the last few years, according to FBI figures in 1993?

    Murder _____

    Burglary _____

Rape _____

Aggravated assault _____

Robbery _____

10. What proportion of the violent crime in the United States is committed by women?

1 percent _____

10 percent _____

20 percent _____

30 percent _____

40 percent _____

11. What is the most common cause of arrest in America?

12. A peculiar activity, once legal but now illegal in many parts of the country, often precedes a shooting. It starts with *S*. Name the activity.

13. In the United Kingdom, the murder rate is 1.3 per 100,000. In the rest of Europe, it is between 2 and 5 per 100,000. What is the murder rate in America?

14. How many U.S. murder victims knew their killers?

One in ten _____

Two in ten _____

Five in ten _____

Eight in ten _____

15. What is the most conspicuous legal form of killing in the United States—at least in thirty-six states?

16. What is a square grouper? Where is the term used?

17. Rank these causes of death of the 12,928 police officers listed on the National Memorial in Washington, D.C., with most frequent cause first:

Shootings _____

Auto accidents _____

Motorcycle accidents _____

Being struck by a vehicle _____

18. How many more drug offenders were sent to state prisons in 1990 than were sent in 1980?

About the same number _____

Half again as many _____

Twice as many _____

Ten times as many _____

Twelve times as many _____

19. Match the cost of raw materials with the method of human execution:

Hanging	$250
Electrocution	Negligible
Lethal injection	31 cents
Gas chamber	$600 to $700

20. Which of the following American white-supremacist groups is thought to be the most violent?

Ku Klux Klan _____

Aryan Nation _____

White Aryan Resistance _____

Skinheads _____

American Nazi Party _____

21. Twenty-three thousand Americans were murdered in 1992. One third of the women murdered were killed by their _____ or their _____ .

22. Where in the United States is the highest rate of violent crime?

23. Between 1977 and early 1993, 191 serial killers had been identified in the United States, with 1,007 known victims. How many serial killers are believed to be operating in the United States at any given time?

About 25 _____

40 to 60 _____

About 300 _____

24. If you wanted to use a drug, and you wanted to be a criminal, and you didn't want to get caught, what drug should you use?

25. Where in the United States is the crime rate highest?

Miami _____

Palo Alto _____

Newark _____

Chicago _____

New York _____

Los Angeles _____

Crime and Punishment

1. Handguns.

2. Seventy-eight percent of rape victims know their attacker.

3. Seventy-eight percent of the violent crimes against whites are committed by whites.

4. Eighty-four percent of the violent crimes against blacks are committed by blacks.

5. Twenty percent of criminals—professionals who keep repeating criminal behavior—commit 80 percent of crimes.

6. Sixty-seven percent of murders produce an arrest; 57 percent of aggravated assaults; 52 percent of rapes; 24 percent of robberies; and 13 percent of burglaries.

7. Most burglaries are committed in the daytime.

8. A serial killer kills, usually one victim at a time, over an extended time with cooling-off periods in between. Christopher Wilder and Ted Bundy are examples. A mass murderer kills a lot of people in a single event. Richard Speck, who murdered eight Chicago nurses, and all those "disgruntled employees" who return to shoot their colleagues are examples. The United States appears to have many more of both kinds of killers than any other country.

9. Only burglary has not increased.

10. Just 1 percent of the nation's violent crime is committed by women.

11. Drunk driving is the most common cause of arrest.

12. Stalking is the activity, and it often precedes a murder. This is a frequent problem for women whose intimate relationships have gone bad. The man cannot let go, he follows the woman, sends her notes, leaves messages on her telephone machine, confronts her at work and on dates, hides in the attic with a gun, defaces old photos, destroys clothing or cars, kills a pet—and sometimes, he kills her. By the early 1990s, thirty-three states had passed antistalker laws that make it possible for police to intervene before the violence gets out of hand.

13. The U.S. murder rate is 10 per 100,000.

14. Eight in ten murder victims knew their killers.

15. Aside from self-defense and police work, executions are the most obvious legal form of killing in the thirty-six states that permit them. Thirteen thousand Americans have been executed here since colonial times, most of them in the twentieth century. Three thousand persons are on Death Row to be killed by gas, hanging, shooting, or lethal injection. According to Amnesty International, only three countries executed more than the United States did in 1991: Iran, China, and the former Soviet Union. All three are well known for their insensitivity to human rights. In addition, the

United States is one of only seven countries that execute people who are under eighteen at the time of their crime. Jack Greenberg, former director of the NAACP Legal Defense Fund, told *The New York Times:* "Virtually all executions have taken place in former slaveholding, segregating states and have been carried out with disproportionate frequency against blacks who kill whites; since 1932, when records were first kept, only one white man who killed a black has been executed." The death penalty was moving north and west, however. While there are 20,000 homicides a year, only 200 murderers are chosen to be executed, and somewhere between 10 and 30 are actually executed. The average executee spends eight years waiting to be taken to the chamber. A selection process this arbitrary can hardly act as a deterrent, though it succeeds in coarsening American life. (See Answer 19.)

16. An ordinary grouper is a fish found plentifully in the Caribbean and the Florida Keys. "Square grouper" is Keys and island slang for a bale of marijuana, often jettisoned by smugglers trying to outrun cops and sometimes found by innocent but alert and vigilant fishermen. A square grouper found still dry and potent within its skins of plastic wrap can be worth a considerable amount of money—$100-plus an ounce on the street in Miami.

17. These are in the correct order. Nearly half the officers listed (6,086) were killed by gunfire. Any criminal can buy or steal a gun anywhere in America.

18. Twelve times as many drug offenders went to prison in 1990 as ten years previously—without significantly reducing drug-related crime. Studies find treatment is far more effective.

19. Lethal injection is the most expensive method, because it is a three-step process involving use of expensive drugs: tranquilizers, sodium thiopental, sometimes cyanide, and prussic acid. Cyanide alone for the gas chamber costs about $250. To electrocute someone takes only some 31 cents'

worth of electricity if all goes smoothly (which it sometimes does not—a few years ago, a natural sponge in the skull cap of an electric chair was replaced by a cheap artificial sponge from a convenience store; when the switch was pulled, flames shot from the victim's head). The cost of hanging is even less, since the price of a thick rope can be amortized over any number of uses. It is far cheaper, by the way, to keep a convict in prison for life (about $600,000 for forty years) than to execute one (which costs an average of $2–3 million for legal work, trials, appeals, reviews, security, and sentencing). (See Answer 15.)

20. According to a study by the Anti-Defamation League of B'nai B'rith, the skinheads, under such names as American Front, SS of America, Aryan Resistance League, and Northern Hammerskins, are the most violent. They target minorities, homosexuals, and homeless people across the country.

21. A third of murdered women are killed by a husband or boyfriend.

22. The U.S. Virgin Islands has a per capita rate of violent crime three times the national average.

23. Some experts guess there are only about 25. Others think the figure is much higher. An FBI spokesperson put it at 50 to 100. Crime journalist Jack Olsen says there may be as many as 300 serial killers at work at any given time. Because of the melodrama of their acts, and the pleasurable little horror-movie thrill of fear we experience in thinking about them, serial killers get much more attention than they deserve.

24. Choose heroin, says one sociologist, Paul Cromwell of the University of Miami, who studied burglars. "Burglars on a depressant drug are less likely to take a serious risk, while users of cocaine would take more risks. That flies in the face of popular wisdom that heroin users are dangerous, but actually they're less dangerous. Heroin addicts are the best criminals. They don't get caught."

25. The U.S. state with the highest crime rate in late 1993 was Florida, and the city with the highest crime rate was Miami, home of one of your coauthors, whose lawn mower was stolen last week. Dade County's serious-crime rate was more than double the national average for 1992. There were 12 serious crimes for every 100 residents. San Antonio was second, followed by Jacksonville, Florida; Palm Beach County, Florida; Fresno, California; Little Rock; Broward County, Florida; New Orleans; Fort Worth-Arlington, Texas; and New York.

Money

America's gray-and-green dollar bill is the de facto world currency. We like it, and so does everyone else. But money is merely a convenient means of sorting, exchanging, measuring, shipping, and storing wealth. Money is not wealth, it merely represents wealth.

In geographic terms, money flows across the landscape like water, draining out of some places and collecting in others according to the influence of forces only economists claim to fully understand. How do we measure money's presence or absence? Why do some regions have more money than others? How do we place monetary value on environmental quality? How do we decide who pays? Can we justify taxing someone in Georgia for maintenance of a national park in Montana? This section is about how we exchange value across time and space by the use of money.

1. Which president appears on the $1 bill? (No peeking.)

2. What are your chances of getting the correct change at an American fast-food shop?

3. The largest portion of the value of U.S. currency in circulation is in which denomination?

$1 bills _____

$2 bills _____

$3 bills _____

$5 bills _____

$20 bills _____

$50 bills _____

$100 bills _____

$1,000 bills _____

4. Where does the word "dollar" originate?

5. According to the U.S. Department of Agriculture, you can feed a family of four nutritionally on $89 per week, with a low-cost plan, or moderately well for about $109. What would it cost on a liberally priced plan?

6. Italians spend 18.8 percent of their income on food. The French spend 15.9 percent. The Greeks spend 31.3 percent. What proportion of our income do we Americans spend on food?

7. In a 1993 study, the average American believed doctors earn about $100,000 a year. We believed they *should* earn about $80,000. What does the average doctor actually earn?

8. Match the salaries for full professors at four-year, liberal arts colleges according to whether the college is public, private or church-affiliated:

$48,600   Church-affiliated

$54,620   Private

$43,210   Public

9. How does the average American private-company pension compare with the average military pension?
The private pension is about half the military pension. _____
The two pensions are about the same, on average. _____
The military pension is about half the private pension. _____

10. In 1992, the Pentagon wasted which amount on unnecessary purchases because of inventory errors?

$40,000 _____

$400,000 _____

$4,000,000 _____

$4,000,000,000 _____

$40,000,000,000 _____

11. In 1992, how much did the average U.S. family pay in federal, state, and local taxes?

About $2,000 _____

About $8,000 _____

About $12,000 _____

About $16,000 _____

About $27,000 _____

12. What proportion of all American households have assets of $1 million or more?

About 1 percent _____

About 4 percent _____

About 11 percent _____

About 16 percent _____

About 7 percent _____

About 10 percent _____

13. Why would you say big corporations are shedding "real" jobs and hiring out as much of their work as possible to part-time and temporary employees?

14. Of all the money Americans give to charity, what proportion would you say is given by individuals, and what proportion by corporations?

83 percent by individuals, 5 percent by corporations _____

51 percent by individuals, 48 percent by corporations _____

46 percent by individuals, 52 percent by corporations _____

7 percent by individuals, 92 percent by corporations _____

15. Say you and your spouse earn $50,000 a year. You have a child. How much should you plan to spend to rear that child to the age of twenty-two?

16. On what collegiate necessity do students now spend more than they spend on textbooks?

17. Where is the top-grossing Radio Shack store in the country?

18. What proportion of small businesses fail in the first five years?

One tenth _____

One third _____

One fourth _____

One half _____

Three fourths _____

19. To die "intestate" is to die without leaving a will. You can imagine what a mess that can create for survivors. How many Americans die intestate?

About 5 percent _____

About 10 percent _____

About 22 percent _____

About 70 percent _____

20. One in five American kids lives in poverty, according to the Children's Defense Fund. Poverty is high among children and low among older Americans. Between 1959 and 1989, the poverty rate for those over sixty-five fell from 35 percent to 11 percent, while for children under eighteen, it rose from 15 percent in 1970 to 22 percent in 1991. Can you suggest any political reason for this disparity?

21. When President Clinton was putting together his health-care plan, did prescription drugs cost the consumer more in the United States or in Canada?

22. At what age does the average American begin saving for retirement?

Twenty-one _____

Thirty-two _____

Forty-three _____

Fifty-six _____

23. How many Americans owned shares in mutual funds in 1992?

One in ten _____

One in five _____

One in four _____

One in two _____

24. Child care is a great responsibility and a heavy expense. But some of us can afford it better than others. If your household income is over $50,000, for instance, you'd spend only 6 percent of your income on child care. What percentage would you spend if you earned under $15,000?

25. Try to match the average weekly salary (in 1991) with the occupation:

Auto mechanic	$ 218
Child-care worker	537
Computer programmer	466
Economist	580
Editor/reporter	385
Farm worker	132
Lawyer	662
Letter carrier	1,008
Social worker	593
Teacher	239
Waitress/waiter	732

26. How much does the average American school-teacher spend, of his or her own money, on teaching materials each year?

27. Where in the United States do citizens earn one half the per capita income of the poorest state, Mississippi?

28. In which part of the United States is it estimated that half the residents receive food stamps?

29. What and where is the building on the back of the U.S. nickel?

30. What and where is the building on the back of the U.S. $10 bill?

31. What and where is the building on the back of the U.S. $20 bill?

32. How many American banks can you name that were, in 1991, among the world's twenty-five largest?

33. In 1978, the Internal Revenue Service got returns from 8,964 persons reporting adjusted gross incomes of more than a million bucks. We know the rich got richer during the 1980s, but were there more of them, or fewer of them by 1990?

34. The average corporate chairman earned thirty-five times more than a typical worker in 1980. Did this proportion grow larger, shrink, or stay about the same during the 1980s?

35. In the early 1990s, how many Americans had jobs paying more than $50,000 a year?

About 100,000 _____

About 1.6 million _____

About 11 million _____

36. Are there more mutual funds or more stocks on the New York Stock Exchange?

37. Unemployment benefits, in 1992, were only $154 a week in Nebraska and South Dakota, the two lowest-paying states. Massachusetts paid the most—how much?

38. Twenty percent of Americans get most of their household finance information from banks, saving and loans, or credit unions. But even more—22 percent—get their financial dope from another source. Can you name it?

39. Forty percent of Americans say they balance their checkbook each time they write a check. What proportion say they balance it each time they receive a bank statement?

40. How do men's and women's views of money differ?

41. Who made more money in the three years after his death than during his entire famous and productive career?

42. In what foreign country do U.S. residents have the most money invested?

United Kingdom _____

Canada _____

Saudi Arabia _____

Japan _____

43. Investors from what foreign country own more of the United States than any other?

United Kingdom _____

Canada _____

Japan _____

Saudi Arabia _____

44. If you discovered gold in your backyard—pretty likely, right?—and dug up enough to make a pair of wedding bands for you and a spouse, how big a hole would be left in the yard?

45. How much money do children between the ages of four and twelve control each year?

46. Americans aged between twenty-five and forty-four are saving money at a pace that will provide them with (10 percent; 20 percent; 30 percent; 60 percent; 80 percent; 90 percent) of what they will need upon retirement to live at their current level. Choose one.

Money

1. Washington. (One in five Americans cannot answer this question correctly.)

2. About fifty-fifty.

3. $100 bills, a large percentage of which are outside U.S. borders.

4. "Dollar" comes from *Joachimsthaler,* which referred to large silver coins minted in Joachimsthal in the former Austrian empire. The town, now called Jachymov, is today in the northwestern Czech Republic.

5. About $134.

6. Americans spend about 9.8 percent on food.

7. The average physician in private practice earned about $192,000 in 1991, after expenses (or six times more than the average American full-time worker). (See Work.)

8. Full professors at public colleges earn an average of $48,600, at private colleges $54,620, and at church-related colleges $43,210. On campuses with doctoral programs, they do much better, earning an average of $63,250, $80,280, and $70,770, respectively. The lower-ranked associate professors, assistant professors, and instructors earn less, of course.

9. American private-company pensions are half of military pensions, even though military personnel work—or "serve"—only about half as long before retiring.

10. The correct answer, according to the Government Accounting Office, is $40,000,000,000 ($40 billion). This was in a single year, remember.

11. The average American family paid $16,110 in taxes in 1992.

12. In 1992, 3,683,000 households in America, or 3.9 percent, had a total net worth of over $1 million. Somewhat fewer—only 3,234,000—had total investable assets of over a million. ("Investable assets" do not include such nonliquid items as real estate and collectibles.)

13. Because it's an excellent deal for them, if not such a great deal for the workers. On average, a part-time worker is paid 60 percent what a full-time worker would get for doing the same job. And the employer need not pay for health benefits, sick leave, or vacation time. For the worker in these jobs, there is no job training, no advancement, no career ladder to climb—only a rut and a paycheck. By the early 1990s, companies were hiring three temps or part-time workers for every full-time worker.

14. Only 5 percent of charitable contributions in America come from corporations. If you have another impression, it is because companies skillfully make a big public-relations splash with every drop in the bucket.

15. It will cost about $265,000 to rear the child to age twenty-two. Plus college costs.

16. Beer, wine, and spirits.

17. In Miami, Florida, which also has the second-highest-grossing Radio Shack store. "It's the Latins," said Arnold Muller, manager of the top shop. "They're 50 percent of our business. They fly into Miami and buy laptops, cellular phones, satellite antennas, and all the new, trendy electronic things."

18. Half of small businesses fail in the first five years. (And plenty of others fail after that.)

19. Seventy percent of Americans die without leaving a will.

20. Both children and old people rely heavily on the government for their income. But old people organize, vote, and lobby their representatives on their own behalf, while children do not.

21. Prescription drugs cost 32 percent more in the United States than in Canada.

22. The average American begins saving for retirement at age thirty-two.

23. One in four Americans owned shares in mutual funds, which are managed by professionals and enable investors to spread their risk among many stocks and bonds. And low interest rates on savings and certificates of deposit were driving more and more investors into the funds—so many that observers wondered if there would be enough money to redeem their purchases when the market dipped and the "naïve" investors panicked and sold. (By the time you read this, we may know how it turned out.)

24. Families—or more likely, single parents—who earn under $15,000 a year must spend 23 percent of their income on child care just so they can afford to work.

25. Auto mechanics earned $385, child-care workers $132, computer programmers $662, economists $732, editor/reporters $593, farm workers $239, lawyers $1,008, letter carriers $580, social workers $466, elementary school teachers $537, waiters/waitresses $218.

26. The average teacher spends $500 of his or her own money on teaching materials.

27. Puerto Rico.

28. Puerto Rico.

29. The nickel building is Monticello, the Virginia home of Thomas Jefferson outside Charlottesville.

30. The $10 building is the Treasury Building in Washington, D.C.

31. The $20 building is the White House in Washington, D.C.

32. There were none.

33. A lot more. By 1990, 60,667 Americans were reporting million-dollar incomes. At the same time, 26 percent more American families fell below the poverty line.

34. By 1990, the average CEO earned 135 times more than the average worker.

35. Nearly 11 million Americans were paid $50,000 or more. More than 1.6 million got $100,000.

36. There are now more American-run mutual funds—some 3,578 of them—than stocks. These have included, as Joe Queenan pointed out in *The New York Times,* an astrologer's fund, a Hawaiian fund, a Catholic bond fund, socially conscious funds, contrarian and hedge funds, single-country funds, single-state funds, single-industry funds. Most of them underperformed the Standard & Poor's 500.

37. Unemployment benefits in Massachusetts paid $444 a week.

38. Twenty-two percent got their financial info from a family member or relative.

39. Sixty percent claim to balance their checkbook after getting a statement.

40. According to some psychologists, women see money as a means of providing and caring, while men see it as a symbol of self-worth.

41. Elvis Presley, who died (some say) at forty-two. Is this why some think he's still alive?

42. Canada.

43. The United Kingdom, by far; its investors are well ahead of the Japanese.

44. The hole would be 6 feet by 6 feet, and 10 feet deep, according to the Worldwatch Institute. That's the amount of soil and rubble removed by U.S. miners to produce, from typical ore, gold enough for two rings. Naturally, there's no gold in *your* yard. Probably.

45. About $15 billion a year, according to James McNeal, professor of marketing at Texas A&M University, and growing by about 20 percent a year. "Believe me," McNeal says, "their income is increasing at a much more rapid rate than yours and mine."

46. About 33 percent. How do you know how much to shoot for? One rule of thumb is that, if you have a pension plan, shoot for a nest egg of two and a half to six times your salary. If you have no pension plan, aim for four to eight times your peak earnings. How we're supposed to go about this is our problem.

Moving Around

Transportation binds a city, a region, a nation, a world together. The links include not only automobiles, horses and carts, trucks and airplanes, but communication in all its forms from the spoken word to telephones, radios, television sets, and now faxes, computer modems, and satellite uplinks.

Movement and communication comprise one of geography's greatest themes. See how much you know about them.

1. What proportion of American households own a bicycle?

13 percent _____

27 percent _____

47 percent _____

69 percent _____

2. How many cars are there in the United States?

3. What kind of car did the James Dean character drive in *Rebel Without a Cause?*

4. How many pairs of shoes does the average American woman buy in a year?

5. What percent of new cars bought in America are air-conditioned?

6. Why are taxicabs yellow?

7. About how many parts are in the average American car?

About 1,000 _____

About 5,000 _____

About 10,000 _____

About 20,000 _____

8. On which two days are the most discount air fares available? On which two days are the most expensive flights?

9. What aircraft have the fewest restrooms per passenger?

767s _____

MD-80s _____

757s _____

737s _____

727s _____

10. Where was the mountain bike invented?

11. Has the number of mobile homes in the United States increased or decreased since 1970?

12. What proportion of domestic mail sent through the U.S. Postal Service consists of personal letters?

4.5 percent _____

14.5 percent _____

24.5 percent _____

44.5 percent _____

67.5 percent _____

13. What proportion of all new cars sold in the United States are foreign?

5 percent _____

10 percent _____

25 percent _____

40 percent _____

45 percent _____

60 percent _____

Minneapolis _____

Washington, D.C. _____

Jersey City _____

14. In one state, fully half the cars sold are of foreign brands. Name the state.

15. In the United States today, a train will collide with a car or truck about every ninety minutes. How many Americans were killed in 1991 in vehicle-train accidents?

About 150 _____

About 250 _____

About 600 _____

About 1,000 _____

About 2,000 _____

16. Which interstate highway is the longest?

I-5 _____

I-60 _____

I-95 _____

I-90 _____

I-75 _____

I-80 _____

17. Which U.S. highway goes from Maine to Key West?

18. What was a "prairie schooner"?

19. The first subway system in the United States was built in 1897—in which city?

Philadelphia _____

New York _____

Boston _____

20. Of all mass transit vehicles in the United States, which is the most commonly used?

21. Are deaths from motor vehicle accidents increasing or decreasing in America?

22. What is the most economical form of travel vacation?

23. In 1983, the average new American car had about eight defects when it rolled off the assembly line. About how many did one have ten years later?

24. You may have noticed that interstate highways with even numbers run east-west, while interstates with odd numbers run north-south. That's your hint. Now, match the interstate highway with the trio of cities it joins.

New Orleans, Memphis, Chicago	I-75
Miami, Washington, D.C., Boston	I-40
San Antonio, Oklahoma City, Kansas City	I-70
Albuquerque, Denver, Cheyenne	I-80
Las Vegas, Salt Lake City, Pocatello	I-90
San Diego, Sacramento, Seattle	I-10
Mobile, Houston, Tucson	I-5
Spokane, Chicago, Cleveland	I-15
San Francisco, Omaha, New York City	I-25
Pittsburgh, Indianapolis, Denver	I-35
Nashville, Little Rock, Flagstaff	I-95
Tampa, Knoxville, Cincinnati	I-55

25. What can you tell about U.S. Highway 66—also known as Route 66? Younger folk may have to quiz their elders on this one.

26. Which is larger—Denver's new airport or the island of Manhattan?

27. What percentage of America's passenger miles traveled were on Amtrak, the nation's passenger railway, in 1991?

28. What is the tallest bridge in the United States?

29. Many of the country's most evocative place names have only three letters and only make sense if you happen to know what they stand for: airports. Here are some familiar three-letter places. Name them:

MIA _____

JFK _____

LAX _____

ORD _____

SFO _____

ATL _____

DFW _____

30. Here are more difficult three-letter places. Can you name any of them?

EWR _____

BNA _____

DCA _____

MCO _____

MCI _____

CVG _____

31. What car brand sold so well in the United States in the late 1970s and 1980s that dealers later admitted bribing factory executives to send them more?

32. How was the expansion of railroads across America financed?

33. As you know, Robert Fulton built the first successful steamboat, the *Clermont*, in 1807. Where was the *Clermont*'s first route?

34. What does the Erie Canal connect?

35. During the cowboy era that peaked in the 1870s, the longest cattle trail ran from Brownsville, Texas, to Fort Buford, North Dakota. It was called the _____ Trail. Fill in the blank.

36. What do the Soo Canals connect?

37. The division of the United States into time zones was formalized in 1883. What were these divisions based upon?

38. How long does it take Amtrak's *Sunset Limited* train to travel from Miami to Los Angeles?

39. What American car, described by *Time* magazine as resembling "an Oldsmobile sucking a lemon," was born in September 1957 and died only twenty-six months later, having sold only 110,000?

40. The average annual cost of owning and maintaining an automobile in the United States is $ _____ , or _____ percent of the typical household's disposable income. Fill in the blanks.

41. How many trees are needed to absorb the carbon dioxide emissions of one American car for a year?

42. Which state has the most covered bridges?

New Hampshire _____

Vermont _____

Pennsylvania _____

Connecticut _____

Georgia _____

Virginia _____

Maine _____

43. The famous Pony Express existed in 1860–61, connecting what two cities?

44. What animal ran for the Pony Express?

45. How much horsepower does a horse have?

46. Before European colonization of the United States, Indian trails were a common feature in the East. Why didn't the auto highways later follow the same routes?

47. Kill Devil Hill. You've probably heard of it, but can you place it?

48. Where is the farthest point inland that an ocean-going ship can penetrate the United States?

49. Before the Civil War, a major highway in the United States was the National Road. Where was the National Road and what did it connect?

50. One city provided most of the locomotive engines that powered America's railroad industry in the 1800s. Which city?

Newark _____

Paterson _____

Hartford _____

Birmingham _____

Portland _____

51. Who were the first American crusaders for well-paved roads?

52. What do these transportation-related initials stand for?

PA _____

EA _____

PI _____

53. In the transportation geography of nineteenth-century America, what were the "Siamese Twins of Commerce"?

54. The song on the first "gold record"—representing sales of 1 million copies—was about a train. Which train? Can you name the artist?

55. Something you might think unusual happens to about 1 million of the cars that become total losses every year in auto accidents. Guess what?

56. Do air bags in automobiles save lives?

57. The German zeppelin *Hindenburg* exploded and crashed in 1937, killing thirty-six and ending the dream of commercial lighter-than-air travel in America. Where did it crash?

58. In number of passengers using the terminals, what were the world's three busiest airports in the early 1990s? (Hint: They're all in the United States.)

59. One of the costs of owning a car is depreciation. Except in special situations that don't apply to most of us, such as ownership of a classic Auburn boattail roadster, a car loses value every minute. In practice, then, are you better off in America to trade your car every year, or every two years?

60. What is the toughest airport in the United States to land in, according to pilots?

61. What two cities were linked by the world's first regularly scheduled passenger airline service?

62. What state has the highest motor vehicle death rate?

63. Can you match these cities with their area codes?

Los Angeles	612
San Francisco	602
Seattle	305
Phoenix	504
Denver	312
Dallas	213
New Orleans	212
St. Louis	303

Minneapolis	206
Chicago	415
New York	617
Miami	412
Pittsburgh	404
Atlanta	314
Boston	214

64. In what state does one in ten residents have a pilot's license?

65. Where was the first television network in the United States?

66. Where is the highest point in the U.S. communication network?

Answers

Moving Around

1. Forty-seven percent of households own at least one bike.

2. There are 123.3 million cars in America, about one for every two of us.

3. A chopped and channeled '49 Mercury. Black, as we recall.

4. The average woman buys 5.6 pairs of shoes in a year.

5. Ninety percent of new cars in America come with air-conditioning.

6. Taxis are yellow because John D. Hertz, who owned a taxi company in Chicago many years ago, commissioned a study by the University of Chicago to determine which color was most visible from a distance. Yellow won. So did Hertz: He also founded a well-known and very profitable car-rental firm.

7. The average car contains 20,000 individual parts.

8. More discount fares are usually available on Tuesdays and Wednesdays. Mondays and Fridays are the most expensive, especially during the rush hours of evening and early morning.

9. The 737s and 757s have the fewest restrooms. They are also the most cramped. The roomiest planes are the 767s and MD-80s.

10. The mountain bike was invented in Marin County, California, in the early 1970s, and first mass produced there ten years later. They were used for zipping up and down the trails of Mount Tamalpias.

11. The number of mobile homes has increased by 230 percent since 1970.

12. Only 4.5 percent of the U.S. mail consists of personal letters.

13. In the early 1990s, 20 percent to 25 percent of cars bought in the United States were foreign makes.

14. California.

15. In 1991, about 600 Americans were killed in more than 5,300 vehicle-train crashes. Some of these, though not listed as such, were suicides—despondent folk driving into the path of a train in order to end it all. Many train-crew members have had to seek therapy or even leave the job because of the memories of what they've seen on the tracks and their inability to prevent tragedy.

16. Interstate-90, which runs 3,107 miles from Boston to Seattle, is the longest.

17. U.S. 1.

18. Prairie schooner was the popular name for the covered wagons pioneers used to cross the Great Plains. Before the land route opened, travelers could reach the West Coast only by sea, aboard schooners with sails.

19. The first subway was in Boston.

20. The bus.

21. In 1992, fewer Americans died in auto accidents than in any year since 1962, partly because of im-

proved cars and improved driving habits encouraged by stronger laws. There were 6.2 million crashes and 39,000 deaths in 1992, down 5 percent from the year before. Twenty years ago, forty-five people died for every billion miles driven, versus only twenty-two people today. We could do even better, experts say, if fewer drinkers drove and everyone wore seat belts.

22. The recreational vehicle, because lodging goes right along with you. Some 25 million of us own or rent rec-vees, and there are 8.5 million of them in use. They can cost anywhere from $2,000 for a little tentlike folding camper-trailer to $63,000 for a fancy motor home. Across America, 20,000 campgrounds welcome rec-vee drivers, 44 percent of whom are aged fifty-five and up. But younger folk use them too: 39 percent are between thirty-five and fifty-four.

23. By 1993, defects in American-built cars were down to an average of 1.4 per car.

24. New Orleans, Memphis, and Chicago are connected by I-55; Miami, Washington, D.C., and Boston by I-95; San Antonio, Oklahoma City, and Kansas City by I-35; Albuquerque, Denver, and Cheyenne by I-25; Las Vegas, Salt Lake City, and Pocatello by I-15; San Diego, Sacramento, and Seattle by I-5; Mobile, Houston, and Tucson by I-10; Spokane, Chicago, and Cleveland by I-90; San Francisco, Omaha, and New York by I-80; Pittsburgh, Indianapolis, and Denver by I-70; Nashville, Little Rock, and Flagstaff by I-40; and Tampa, Knoxville, and Cincinnati by I-75.

25. Route 66 was a culturally significant highway across the American West. It ran from Chicago to Los Angeles, a distance of 2,448 miles, and in the 1930s, it carried poor migrants from Oklahoma and nearby states to California, land of opportunity. Another generation followed it to the defense plants near Los Angeles in the decades after World War II. Bobby Troup wrote a famous song about it in 1946, and *Route 66* was the title of a popular television series in the early 1960s. Route 66 was officially decertified in 1984. Today, about 2,000 miles of the original highway survive as frontage roads, interstate loops, and renamed local roads. The longest remaining stretch is 157.8

miles in northern Arizona. It is the longest state historic monument in the United States.

26. Denver's new airport covers 53 square miles, making it about twice the size of Manhattan.

27. About 1 percent. Seventy-nine percent of the nation's passenger miles were in automobiles.

28. The Golden Gate Bridge, connecting San Francisco with its northern suburbs, is 746 feet high, the highest in the country.

29. These are Miami International; John F. Kennedy in New York; Los Angeles International; O'Hare International in Chicago (whose original name was Orchard Field); San Francisco International; William B. Hartsfield/Atlanta International; and Dallas-Fort Worth International.

30. These are the airports in Newark, Nashville, Arlington (Washington National Airport), Orlando, Kansas City, and Covington, Kentucky (Cincinnati/Northern Kentucky Airport).

31. Honda. Even after the Japanese company built a factory in Marysville, Ohio, and began producing more cars there, they found it hard to meet demand for Civics, Accords, and Preludes.

32. By real estate development companies, with a crucial subsidy from the government: free land. The firms were simply handed enormous land grants along their rights-of-way, as well as other lands rather distant from their rights-of-way. These lands were then developed as destinations to encourage settlement, or sold at a profit to settlers. The corporate descendants of these successful examples of "private enterprise" still own huge amounts of the American West.

33. The *Clermont* proved its worth along the Hudson River route from New York City to Albany. In 1820, a steamship named S.S. *Robert Fulton* became the first vessel to steam from New York to Havana, Cuba.

34. The Erie Canal connects the Mohawk River valley with Lake Erie, thus bringing Albany and Buffalo into prominence and indirectly guaranteeing the trade dominance of New York City.

35. Western.

36. Lake Superior and Lake Huron. Before they were built around the falls at Sault Sainte Marie (pronounced "Soo Saint Marie"), Michigan, in 1855, ships were unloaded at the falls and cargoes carried overland in wagons and wheelbarrows.

37. Time zones in the United States were based on railroad timetables.

38. The *Sunset Limited*, running three times a week, takes about sixty-five hours to run from Miami to Los Angeles, a distance of some 2,944 miles.

39. The Edsel.

40. Owning and maintaining a car eats up some $3,500 a year, or 12 percent of a typical household's disposable income. Here's the breakdown for one example, a car purchased for $14,000 in 1991, kept for twelve years, and driven 128,500 miles. Depreciation and interest cost $16,300; insurance, $9,050; gas, $7,800; repairs and maintenance, $5,350; tolls and parking, $1,650; taxes, licenses, registration, $1,100; tires, $1,250; oil, $200. Total cost of owning the car over twelve years: $42,700. (And one other cost—see Answer 41.)

41. Two hundred trees.

42. Pennsylvania has the most covered bridges.

43. The Pony Express connected St. Joseph, Missouri, and Sacramento, California.

44. The horse. Not the Shetland pony, of course. "Pony" is the casual term used by horsy folk to designate a handy little workaday steed.

45. Scientists at the University of Massachusetts and Dalhousie University in Nova Scotia recently found that, theoretically, a typical horse weighing 600 kilograms would produce about 24 horsepower.

46. The answer illustrates a point of geography: Everything may be somewhere, but that may not be the right place. The Indian trails were ideally located for their purpose, following mountain ridges and hilltops because the drainage was good, which make for better footing. But ridges and hilltops are less welcoming to wagons and construction machinery. So highways were often placed on the lower, flatter terrain beside rivers, where it was easier to build. This, by and large, is where towns began to grow.

47. Kill Devil Hill is where Orville Wright took off for his twelve-second flight near Kittyhawk, North Carolina, in 1903. The Wright brothers were from Ohio but went to coastal North Carolina for the strong, reliable winds, which they hoped would help them get that contraption into the air. They were right.

48. Duluth, Minnesota, on Lake Superior, via the St. Lawrence Seaway and Lakes Ontario, Erie, and Huron. The farthest inland you can go exclusively by river would be on the Columbia-Willamette River to near Portland, Oregon.

49. The National Road connected Cumberland, Maryland, with Columbus, Ohio, and went on to Vandalia, Illinois.

50. Three of the greatest locomotive works were in Paterson, New Jersey.

51. Bicyclists formed the League of American Wheelmen, and by 1890 they were lobbying mightily for smooth places to ride.

52. Extinct American airlines: Pan American, Eastern Airlines, Piedmont Airlines. Can you add others to the list?

53. The twins were the railroads and the telegraph.

54. The 1942 hit was "Chattanooga Choo Choo" by Glenn Miller.

55. About a million totalled cars are "repaired" and resold, some with unsafe brakes or steering. Since this is illegal, many are shipped to other states and retitled, and dealers lie about their provenance. Some of these cars have crashed, seriously injuring their buyers.

56. Through 1992, air bags had saved some 500 American lives in highway accidents. They've reduced the chance of fatal injury by 23 percent and of severe injury by 68 percent. Air bags are expected to save another 2,400 lives between 1990 and 1995. By 1997, all new passenger cars will be required to have them on both driver and passenger side. They work best when used in conjunction with seatbelts.

57. Lakehurst, New Jersey.

58. Number one was Chicago's O'Hare, number two was Dallas-Fort Worth, and number three was Los Angeles International. John F. Kennedy Airport in New York, however, moved the most tonnage of cargo.

59. According to the U.S. Dept. of Transportation, the driver of an intermediate car who trades it in annually will pay $52,000 in depreciation alone over twelve years. Someone who trades every two years instead will pay only $34,000 in depreciation in the same period.

60. National Airport in Washington, D.C., holds that dubious honor. Eighty-two percent of pilots surveyed said it was the most "challenging." La Guardia in New York was second.

61. Tampa and St. Petersburg, Florida, in 1914.

62. Nevada, with 3.6 deaths per million vehicle miles in 1990. The highest number of auto deaths per capita is in New Mexico.

63. Los Angeles's area code is 213, San Francisco's 415, Seattle's 206, Phoenix's 602, Denver's 303, Dallas's 214, New Orleans's 504, St. Louis's 314, Minneapolis's 612, Chicago's 312, New York's 212, Miami's 305, Atlanta's 404, Boston's 617, and Pittsburgh's 412.

64. Alaska.

65. The DuMont network linked Washington, D.C., and New York City in 1946.

66. A constellation of communication satellites in geostationary orbit is 22,300 miles above our heads.

Cities

Cities are where people first began learning to live closely together, a lesson which remains incomplete. We tend to take the city so much for granted that we may no longer recognize it as one of humankind's greatest inventions. In cities, we share ideas quickly, pool collective wisdom to make better decisions, and preserve a record of past experience upon which to build the future. Cities provide a variety and depth of available work, which invites new population. And that creates and satisfies a desire for culture—the invention and artful dissemination of original ideas.

Now, however, the city-bound view of culture may be changing. With phones and faxes and computer networks and easy transportation, many of us can do city things while living in the distant suburbs or the country.

Meanwhile, the cities still have their museums and concert halls, but civility—even safety—appear to be on the wane.

1. In the 1960s Beatles song "Back in the U.S.S.R.," the first city named is not in the old, broken-up Union of Soviet Socialist Republics, but back in the U.S.A. Which city is that?

2. What is the fastest-growing city in the United States?

3. In 1790, the five largest cities in the United States were Boston, New York, Philadelphia, Baltimore, and Charleston, South Carolina. What do the physical settings of these cities have in common?

4. Many American cities grew up around colonial forts. Name the city that grew up around each of these forts:

Fort Dallas _____

Fort Dearborn _____

Sutter's Fort _____

Fort Conde _____

Fort Lowell _____

Fort McHenry _____

Fort Wayne _____

Fort Hamilton _____

Fort Sumter _____

Fort Duquesne _____

The Alamo _____

Fort Worth _____

5. In the year 2010, the most new jobs will be created in ten of the following eleven American metropolitan areas. Which one does not belong?

Los Angeles _____

Anaheim _____

Phoenix _____

San Diego _____

Dallas _____

Houston _____

Seattle _____

Washington, D.C. (including Fairfax County, Va.) _____

San Jose _____

Fort Lauderdale _____

Pittsburgh _____

6. Of the 4 million highest-paid Americans, half live in a single city. Can you guess which one?

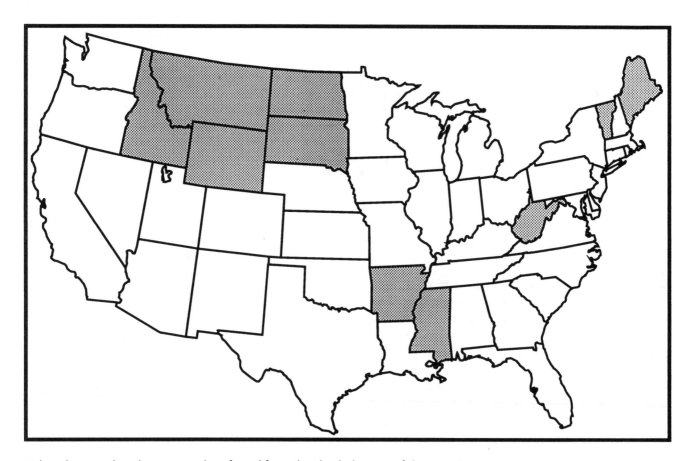

What distinguishes the geography of city life in the shaded states of this map?

Answer: These are the least urbanized states in the United States, according to the 1990 census. Fewer than 40 percent of the people in these states live in metropolitan areas. The *most* urbanized states were California, New York, Massachusetts, New Jersey, Connecticut, Maryland, and Florida. They all had more than 90 percent of their residents living in cities.

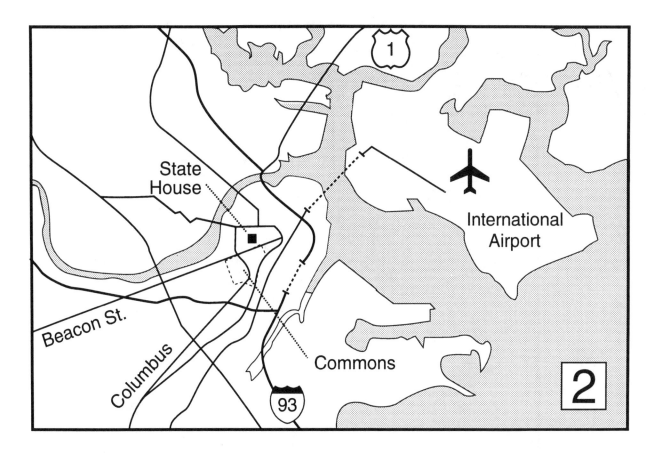

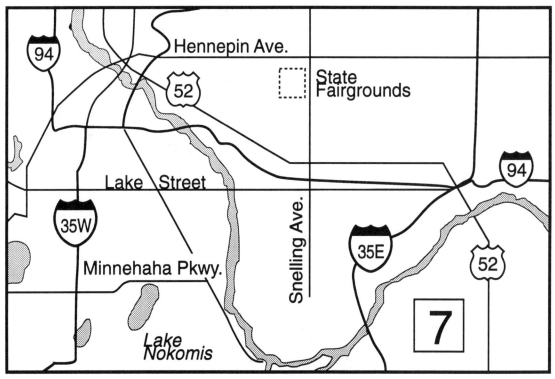

Here are maps of ten American metropolitan areas. Your task is to identify each of them. (Hint: All but one have teams in the National Football League—and the other takes college sports very seriously.)

Answers: 1. Atlanta, 2. Boston, 3. Columbus, 4. Miami, 5. Denver, 6. Houston, 7. Minneapolis/St. Paul, 8. Phoenix, 9. Seattle, 10. Indianapolis.

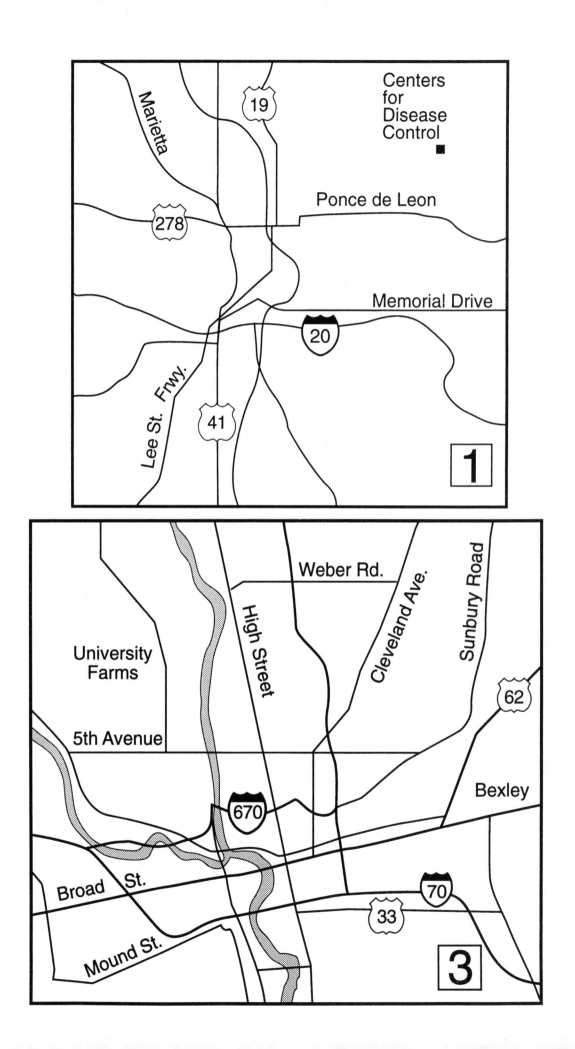

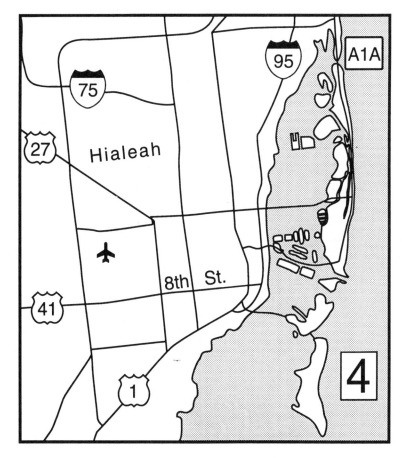

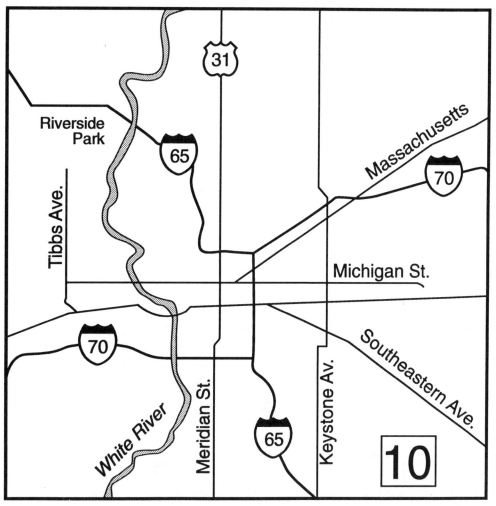

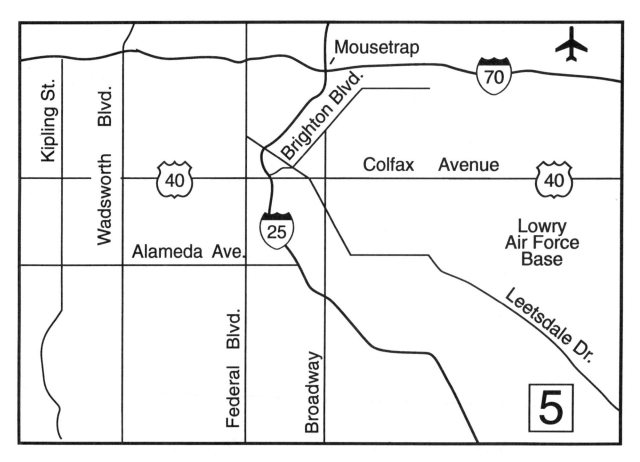

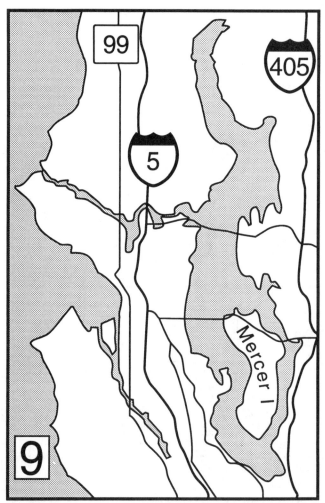

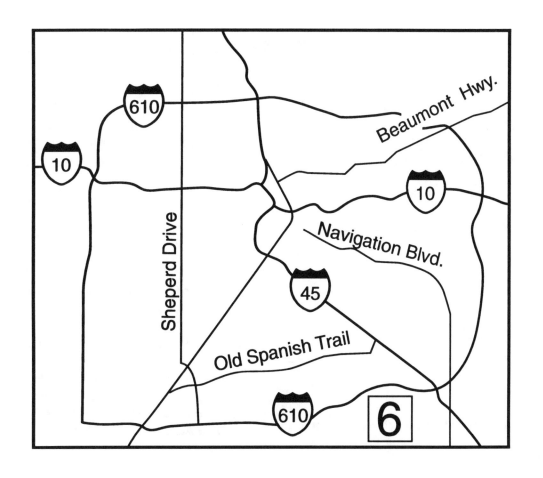

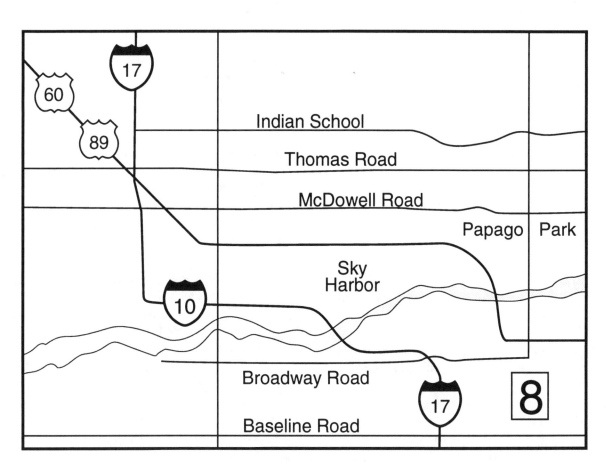

7. In what city does Clark Kent work?

8. Which U.S. city has the highest rate of poverty?

9. Which of the following three cities is farthest west?

Reno _____

Las Vegas _____

Los Angeles _____

10. Which of the following three cities is farthest north?

New York _____

Chicago _____

Denver _____

11. Patience and Fortitude are highly visible characters in New York City. Can you say who they are?

12. Identify the city from its nickname:

The Crescent City _____

The Magic City _____

The Rose City _____

The Windy City _____

The City of Brotherly Love _____

The Queen City _____

The City of Angels _____

13. What city's name means "red stick"?

14. Where were the seven Cities of Cibola?

15. What was the capital city of early Spanish California?

16. Name an American city in which more people receive welfare than attend public schools.

17. Francis Scott Key wrote our unsingable national anthem, the "Star Spangled Banner," while observing the shelling of Fort McHenry in 1814. Where is Fort McHenry?

18. Where is Mardi Gras?

19. What city was designed by Pierre Charles L'Enfant, who was sacked by President George Washington before the project was finished?

20. In what city does Bruce Wayne live?

21. In what city was the major leagues' first night game?

22. What city was made famous by the Parker Brothers in their game Monopoly?

23. One of America's great novels, *The Jungle*, by Upton Sinclair, vividly described the unsanitary and inhumane conditions in stockyards and slaughterhouses, turning many readers into vegetarians overnight, and led to the Pure Food and Drug Act of 1906. In what city was *The Jungle* set?

24. In what American city was the United Nations founded?

25. What city in the United States has the most motel rooms? (Hint: Think of Mickey Mouse.)

26. What is Chicago's "Loop"?

27. What city was the home of the Toughy Gang, the Ralph Sheldon Gang, De Coursey's Gang, and the Capone Gang?

28. What city is the home of McDonald's Corporation?

29. Ernest Hemingway was born near Chicago, but his name later came to be associated with two other American cities, both noted for outdoor recreation—one for fishing, the other for hunting and skiing. Can you name them?

30. What single geographic factor explains the location of most U.S. towns and cities?

31. Why did New York emerge as the largest city in the East?

32. Why did Los Angeles emerge as the largest city in the West?

33. New Amsterdam was New York City's original, ethnic name. Of what U.S. city was San Fernando de Bexar the original, ethnic name?

34. Two American cities were actually planned to be state capitals but are no longer state capitals. Can you name one? Or both?

35. Pronounce these as the residents do:

Biloxi, Mississippi

Versailles, Kentucky

Cairo, Illinois

Peru, Indiana

Lima, Ohio

36. In what town was President William Jefferson Clinton born on August 19, 1946?

37. What city in the United States is largest in land area?

38. On most in-country airline flights, demand for hard liquor has virtually evaporated in the past few years, what with health-conscious passengers choosing wine, beer, nonalcoholic beer, fruit juice, and designer water as their beverage-of-flight. But airlines flying to one city still must carry liquor because people going there still want it. Can you name the city? (Hint: Think of Wayne Newton.)

39. Two parts of Manhattan Island are most densely and highly developed with high-rise buildings. What are these sections called?

40. Which came first in America, the railroad or the city?

41. Which city has Jonathan Raban called "the first big city to which people had fled in order to be closer to nature"?

42. A printed recipe for ice cream first appeared, in 1792, in a cookbook of what city? (Hint: Think cream cheese.)

43. What is a "fall-line" city?

44. What city is the communications center of the country?

45. In which city did Henry Ford build his first auto factory?

46. What city is probably the agricultural capital of the United States?

47. What city is often called the rubber capital of the world?

48. New Orleans, one of the major ports in the world, is situated about how far from the Gulf of Mexico?

49. What city, despite economic difficulties, is probably still the nation's oil capital?

50. In what city was the first washateria? And what in the world is a washateria?

51. Only one major city in the United States can boast of 365 days a year without unhealthful levels of air pollution. Is it Miami, Las Vegas, Detroit, San Francisco, Denver, or Houston?

52. Here is a list of cities. Pick the five most livable, according to a *Money* magazine survey in 1993:

Rochester, Minnesota _____

Madison, Wisconsin _____

Minneapolis-St. Paul _____

Houston _____

Raleigh-Durham, North Carolina _____

Goldsboro, North Carolina _____

Sheboygan, Wisconsin _____

Grand Forks, North Dakota _____

Sioux Falls, South Dakota _____

Austin _____

Seattle _____

Yuma _____

Provo-Orem _____

Salt Lake City-Ogden _____

Denver _____

Milwaukee _____

St. Joseph, Missouri _____

Columbia, Missouri _____

Abilene, Texas _____

Fort Worth-Arlington, Texas _____

53. What two states gave up territory to form the District of Columbia?

54. What U.S. city takes its name from a coastal hamlet in England called St. Bartolph's Town?

55. What is the northernmost city in the United States?

56. Only one state capital is on the Mississippi River. Can you name it?

57. What industrial city, hometown of Andy Warhol, has been known as The Smoky City and "Hell with the Lid Off"?

58. What do Marquette, Michigan, and Joliet, Illinois, have in common?

59. What famous American city is made up of fourteen islands?

60. Residents of what city pay the highest per capita city income taxes?

61. Here is a list of cities. Pick the one chosen as least friendly in the world by subscribers of a well-known travel magazine.

Palm Beach _____

Las Vegas _____

New York _____

Los Angeles _____

Miami _____

62. Where did the Wizard of Oz live?

Cities

1. Miami Beach, Florida.

2. As of the early 1990s, Fresno, California, was the fastest-growing city.

3. Each was situated on a sheltered, navigable river. These cities functioned as ports for the raw-materials exports of the early United States and for import of the manufactured goods the early colony needed.

4. Miami, Chicago, Sacramento, Mobile, Tucson, Baltimore, Detroit, New York, Charleston, Pittsburgh, San Antonio, Fort Worth.

5. Pittsburgh does not belong on this list.

6. New York City.

7. Clark Kent, mild-mannered reporter for the great metropolitan newspaper the *Daily Planet,* lives and works in Metropolis. His alter ego, Superman, the man of steel of DC Comics, was born on the planet Krypton.

8. Detroit has the highest poverty rate, according to the U.S. Census.

9. Reno is at 119 degrees west longitude, Los Angeles is at 118 degrees, and Las Vegas at 115 degrees. Therefore Reno is the westernmost.

10. Chicago is at 41 degrees north latitude, Denver at 39 degrees, and New York at 40 degrees. Therefore Chicago is farthest north.

11. Patience and Fortitude are the stone lions in front of the New York Public Library.

12. New Orleans; Miami; Portland, Oregon; Chicago; Philadelphia; Cincinnati; Los Angeles.

13. Baton Rouge.

14. The Seven Cities are nowhere. They never existed. But the Spanish explorers of modern Arizona and New Mexico in the 1500s seem to have believed they did, having been told as much by many of the local Indians, who used the Spanish greed for gold to get rid of them by describing fabulously wealthy cities in a place just over the horizon. Legends of lost treasure are still a tradition in the American Southwest.

15. Monterey, where many Spanish colonial buildings survive. Most are open to tourists.

16. There may be more such cities, but New York is one.

17. Baltimore, Maryland.

18. Since Mardi Gras, meaning "Fat Tuesday" in French, is a religious rite rather than a place, presumably it is everywhere. The holiday, before the start of austere Lent, is celebrated with superior gusto in New Orleans.

19. Washington, D.C. Andrew Ellicott completed the project.

20. Bruce Wayne—and his alter ego, Batman—lives in Gotham City. "Gotham" is a nickname for New York, of course, but its origins go back to England, where Gotham was a mythical city famed for the endearing befuddlement and foolishness of its inhabitants. There is a place just like it in the Yiddish tradition, where it is known as "Chelm."

21. The major leagues' first night game was held in 1935, at the Redlegs' home, Crosley Field in Cincinnati, Ohio.

22. An unemployed engineer named Charles Darrow invented Monopoly, basing its layout on the streets of Atlantic City, New Jersey. Monopoly has sold 100 million copies in twenty-seven languages, become the most popular game of all time, earned millions for the Parker Brothers game company, and made Atlantic City famous.

23. *The Jungle,* published in 1905, was set in Chicago.

24. The UN was founded in San Francisco, California, in 1945. It moved later to Gotham.

25. Orlando, home of Disney World, the world's most popular tourist destination, has the most hotel / motel rooms.

26. The word refers to Chicago's central business district and derives from the curving route of the city's elevated railway.

27. Chicago in the 1920s.

28. Des Plaines, Illinois, is the home of McDonald's.

29. Key West, Florida, and Ketchum, Idaho.

30. Transportation. In colonial days, it was the port, with its wharves, around which a city grew. Later it was the railroad, with such crucial hubs and junctions as Chicago. Today, new cities are springing up at freeway interchanges at the edge of old railroad towns. These new places depend on trucks.

31. Transportation. New York has the biggest and best natural harbor on the Atlantic seaboard. When the Erie Canal gave New York's port inland access to the Great Lakes, New York's power was certified.

32. Weather, water, and fantasy—not transportation. Los Angeles surged because the sunny, warm weather attracted the early movie industry to Hollywood (from places like New Jersey). It was

sunny and warm because it was largely desert, however, and growth would not have been possible without the large-scale importation of water from northern California and the Colorado River. Most of the growth, though, is based on the Middle-American fantasy of a new, free lifestyle that is imagined to exist on the West Coast. And does, if you believe the tabloids.

33. San Antonio, Texas.

34. Iowa City, Iowa, and Prescott, Arizona.

35. Buh-LUX-ee, Ver-SALES, KAY-ro, PEE-roo, LY-ma.

36. Hope, Arkansas.

37. Jacksonville, Florida.

38. Las Vegas.

39. Wall Street and Midtown.

40. Most eastern U.S. cities were there before the railroad, having originated at good ports and harbors, river crossings, mountain gaps and other trail intersections. The railway station is located like the afterthought it was. In western cities, built generally after or by the railroad, the railway station became a civic focal point. The railroads stimulated quicker city growth and in some of the western towns, street grids were laid out to parallel the rail line, breaking up the typical north-south grid.

41. Seattle, so close to mountains, sea, and forests.

42. Philadelphia.

43. A fall-line city is one situated where the low, flat Coastal Plain meets the gently rolling Piedmont. Examples are Trenton, Richmond, Raleigh, and Baltimore.

44. New York, which leads the country in publishing books, magazines, and newspapers, and is headquarters to major radio and television networks and press services.

45. Detroit.

46. Omaha.

47. Akron.

48. New Orleans is 110 miles up the Mississippi River from the Gulf Mexico.

49. Houston.

50. The U.S.'s first Laundromat, then called a washateria, was built in Fort Worth in 1934.

51. It is San Francisco. Los Angeles is the worst, with 158 days a year of unhealthful levels, followed by Houston with 40, New York with 26, Philadelphia with 24, Washington, D.C., with 16, Detroit with 11, Chicago with 8, Denver with 7, Atlanta with 6, and Boston and Pittsburgh with 4.

52. The first five are the most livable. The rest of the survey's top twenty follow in order.

53. Maryland and Virginia. Virginia took its land back during the Civil War, however, and to this day has not restored it to the capital.

54. Boston.

55. Barrow, Alaska, on the Arctic Ocean, is northernmost.

56. St. Paul.

57. Pittsburgh.

58. They were named for two French explorers of the Great Lakes who traveled together.

59. New York is made of fourteen islands, including Staten Island, Roosevelt Island, and Hart Island in the Bronx, site of Potter's Field, where some 3,000 paupers are buried in mass graves each year.

60. Washington, D.C., residents pay the highest local income taxes, followed by New Yorkers—who also must pay state taxes.

61. Readers of *Condé Nast Traveler* said Miami was the least friendly in the world, followed by Los Angeles, New York, Las Vegas, and Palm Beach in that order. The friendliest were Sydney, Dublin, Auckland, Kauai (actually an island rather than a city), and Victoria, British Columbia.

62. The Wizard of Oz lived at the end of the yellow brick road in the Emerald City, supposedly a place of prosperity and happiness governed by a benevolent despot. Of course, such a city never existed.

Religion and Belief

Often denounced as a godless nation, America probably leads the world in church attendance. America has invented, created, or served as the birthplace of more religions than any other country—and we don't mean only the Pursuit of the Almighty Buck. There are the Seventh Day Adventists, the Church of Jesus Christ of Latter Day Saints (Mormons), the Black Muslims, Scientologists, Jews for Jesus, Jehovah's Witnesses, and any number of smaller sects, a few of which occasionally flare into national prominence, and any of which, for all we know, could suddenly catch on and start gaining adherents by the thousands. We Americans invented the drive-in church and the television evangelist.

Wherever we live, we humans appear to want to believe in something higher, stronger, and wiser than ourselves—perhaps because we recognize our limitations and desperately seek a better idea. Geographers examine religion as it relates to region (as in the Mormons of Utah) or in its cultural and economic contexts (the Bible Belt) or as it alters the landscape (what sort of neighborhoods have the most churches?). They examine pilgrimages and their termination points, sacred places, and images of the world that religions advocate.

Here is a reasonably reverent look at religion and belief in the U.S.A.

1. What proportion of Americans describe themselves as atheists or agnostics?

   About 1 percent _____

   About 4 percent _____

   About 11 percent _____

   About 21 percent _____

2. One in seventeen Americans claims to have had a supernatural experience, except in one state—

where one in three makes that claim. Which state?

3. What do America's Norwegian, Danish, Swedish, Finnish, German, Estonian, and Latvian communities share?

4. "God created man pretty much in his present form at one time within the last 10,000 years." That is a fair summary of the creationist point of view, which rejects the notion of evolution. About what proportion of Americans buy creationism?

   About 17 percent _____

   About 26 percent _____

   About 33 percent _____

   About 47 percent _____

   About 56 percent _____

5. What are the chances that the next person you bump into is a born-again Christian?

6. Where will you find the Amana Society?

7. A religious cult is a group marked by unconventional beliefs, charismatic leadership, and strict control. How many such cults are there in America, would you say?

   About 100 _____

   About 300 _____

   About 500 _____

   About 800 _____

8. Do frequent churchgoers steal postage stamps from the office as frequently as nonchurchgoers?

9. What proportion of Americans say religion is very important in their lives?

10. "Church of God" seems the perfect name for a religion, and a lot of believers seem to agree. About how many denominations in America use this phrase as the basis of their name?

11. What do Hartford, Omaha, Providence, Louisville, and Miami have in common, religiously speaking?

12. In 1750, what was the major religion of:

Virginia? _____

Massachusetts? _____

New York? _____

13. Who was Vernon Howell? In what religion was he a leader?

14. Perhaps America's greatest female religious leader was Mary Baker Eddy. What did she found?

15. One of the country's other great woman preachers was Aimee Semple McPherson. What church did she found, and where?

16. When Utah became the forty-fifth state in 1896, what did its citizens have to do in order to gain admission?

17. Which part of the country has the fewest religious fundamentalists—that is, persons who believe in the literal truth of biblical accounts of the creation?

18. One of the world's most farflung religions is the Baha'i Faith, which upholds the oneness of God and humankind. Where are Baha'i headquarters in the United States, and an impressive monument?

19. Match the proportion of Americans with their religious preference:

Protestant	2 percent
Catholic	56 percent
Jewish	26 percent
Muslim	.5 percent

20. What single religion in the United States has the most adherents?

21. One in four Americans consults the morning horoscope before reading the headlines. A third believe they have been in touch with a dead relative. Believers in the New Age embrace psychokinesis, telepathy, channeling, pyramid power, UFOs. Most of the Branch Davidians who died in Waco, Texas, were college educated and some had advanced degrees. What do these people have in common, according to Dr. Robert L. Park, professor of physics at the University of Maryland, who advanced these examples?

22. What and where is the Crystal Cathedral?

23. Where is the Little Brown Church in the Vale?

24. Where was—and still is—America's first Mardi Gras?

25. What city is the hometown of Jehovah's Witnesses?

26. What are the four main varieties of Judaism in the United States?

27. Mormons (members of the Church of Jesus Christ of Latter Day Saints) believe the Bible as well as the Book of Mormon. So, what does the Book of Mormon tell about?

28. Where would you find the Church of Satan?

29. America's leading theological thinker and scholar may be Dr. Martin Marty, who teaches and researches at a secular university. What university?

30. In which region of the U.S.—East, Midwest, South, or West—did an early 1990s poll show the highest rate of church attendance?

31. In what major American city did the Black Muslims get their start?

32. The Holocaust Museum in Washington, D.C., had more than 500,000 visitors in its first six months. What proportion of these visitors were Jewish?

About 98 percent _____

About 86 percent _____

About 76 percent _____

About 48 percent _____

About 38 percent _____

33. What is the Code of Handsome Lake?

34. What is "Chutzpah on the Chudson"?

Religion and Belief

1. Only 4 percent of Americans consider themselves to be atheists or agnostics. The rest are religious, or just-in-casers. Eighty-two percent say they are Christians (56 percent Protestant, 25 percent Roman Catholic, 1 percent Eastern Orthodox) and 2 percent are Jewish. (The nation with the highest proportion of nonbelievers is Uruguay, with 10 percent.)

2. California.

3. All have spawned at least one variety of Lutheranism in the United States at one time or another.

4. About 47 percent of Americans agree with the creationist position as given, including one fourth of college graduates.

5. One in three.

6. The Amana Society had its origin as the Community of True Inspiration in 1714 in the district of Hesse in Germany. Following a divine inspiration in the 1840s, members trekked to the United States, first to Buffalo, New York, then to eastern Iowa in 1855, where they still are. Their often communal piety led to successful businesses, which are now separate from the church.

7. About 700 to 800 religious cults are promising paradise to the alienated and disenfranchised.

8. Churchgoers, and especially people who belong to small religious groups that meet frequently, are less likely to steal than others, according to a survey of 2,000 workers (if they told the truth).

9. A poll in 1993 found that about 58 percent say religion is very important in their lives, and an additional 29 percent say it is somewhat important. Only 12 percent thought religion was unimportant. But such figures fluctuate, as if the subject were some kind of fad. Only 53 percent said religion was important just five years before, and in 1952, it was 75 percent.

10. At least sixteen groups, spanning many varieties of American Protestantism including some Mennonites and some Seventh Day Adventists, use "Church of God" in their names. There are nearly as many using "Church of Jesus Christ."

11. These cities, and more than a dozen others, are all seats of archbishops of the Roman Catholic Church in the United States.

12. In Virginia, the major religion was Anglican (Church of England). In Massachusetts it was Congregational; in New York (you knew this) it was Dutch Reformed.

13. Better known as David Koresh, Howell was leader of the Waco-based Branch Davidians, who shot four federal agents during a raid that led to a fifty-one-day standoff ending with an FBI assault and a fire in which seventy-two cultists, including nearly a score of children, died. Like Jesus, Koresh died at age thirty-three.

14. Mary Baker Eddy founded the Church of Christ, Scientist. Traveling through New England and New York during the 1860s and 1870s, she spread the idea that physicians were unnecessary.

15. Aimee Semple McPherson founded the International Church of the Foursquare Gospel in 1927 in Los Angeles. Still highly evangelical, its members hope to have 2,000 churches by the year 2000.

16. To get into the United States, Utah citizens had to declare polygamy illegal. They did so, but the practice has remained a quiet tradition among conservative Mormons.

17. The Northeast has the fewest fundamentalists—only 39 percent believe the Bible's account is literally true, while 46 percent nationally believe it is. Even so, 93 percent of Northeasterners believe in God, and 40 percent go to church or temple weekly. The Northeast also has more Muslims than any other portion of the country: one half of 1 percent.

18. Baha'i Faith headquarters is in Wilmette, Illinois.

19. Fifty-six percent of Americans are Protestant, 26 percent Catholic, 2 percent Jewish, and one half of 1 percent Muslim. Islam is the fastest-growing religion in the United States. Muslim leaders claim they have 6 million adherents in the United States, though a study done at the City University of New York found closer to 1.4 million.

20. Roman Catholics make up the single largest group, with 20 million. There are more Protestants, but they are subdivided.

21. They believe in superstition rather than evidence. "What connects all superstitions is the belief that humans occupy a position of central importance in the universe—that everything was constructed with us in mind," says Park.

22. The Crystal Cathedral, in Garden Grove, California, just a few miles south of Disneyland, is a large glass church operated by the rather middle-of-the-road televangelist Robert Schuller. The Christmas pageant features live animals.

23. Near Nashua, Iowa, where it does a big business in marriages.

24. Mobile, Alabama.

25. The Jehovah's Witnesses began in Allegheny City, now part of Pittsburgh, in the 1870s. Their headquarters today are in Brooklyn, New York.

26. Reform, Conservative, Reconstructionist, and Orthodox are the main four.

27. The Book of Mormon tells of Christ's dealings on the American continent from about 600 B.C. to A.D. 421.

28. The Church of Satan is in San Francisco, has operated openly since 1966, and seems to be gaining popularity. Among its precepts are indulgence and vengeance.

29. The University of Chicago, alma mater of one of your authors.

30. The Midwest, where 46 percent reported church attendance during the previous week.

31. The Black Muslims began in Detroit, where in 1930 a seller of silks and other goods named Wallace D. Fard announced himself the "brother from the East." His origins are mysterious, but he had picked up the basics of Islam and a little Arabic. His message of faith, mysticism, and redemption found many hearts. Later the Black Muslim center moved to Chicago, where Fard's first major convert, Elijah Poole (later Elijah Muhammad) expanded the faith.

32. About 38 percent of visitors to the Holocaust Museum were Jewish during the first six months. And 48 percent were Christian (26 percent Protestant and 22 percent Roman Catholic). "We knew this museum had a very solemn and very special message for all Americans," said Miles Lerman, chairman of the U.S. Holocaust Memorial Council.

33. The Code of Handsome Lake is the most potent modern vision of American Indian religion. Handsome Lake (Ca-ne-o-di-yo) was an Iroquois from northern Pennsylvania who, in 1799, had a deathbed vision and resurrection. He preached the "Good Message," an Indian nationalist philosophy that included respect for the Great Spirit and contempt for alcohol. He urged Indians to reject European culture generally and not to sell their land to whites. These teachings, in one form or another, have been basic to most Native American religions ever since.

34. "Chutzpah on the Chudson" is the title of a Jewish radio show in New York (1300 on the AM dial). The station also broadcasts "Ask the Rabbi" and Jewish music. It rests on the seventh day, however, broadcasting Haitian and Latin music from Friday night until Saturday night, the Jewish sabbath. "Chudson," of course, refers to the well-known river west of Manhattan.

Images

Image always has to do with display, show, surface—the impression something makes as well as its implicit reality. Things are not always what they seem, and images are not always truthful to what they seem to represent, although sometimes they may be.

It can be fun to manipulate images. Our clothes say something about how we wish to be seen. We call it fashion. We read "shelter" magazines for help in imagining the ideal place to live. We kid ourselves a little, too. *Purple mountain majesties, amber waves of grain* . . . gives us a little chill, doesn't it? Just the phrases, even though other countries have purple mountains and wheat fields, and even though we've strip-mined some of ours and poisoned much of the fertile soil with pesticides. The image still has a kind of reality for us, not a bad thing.

America is full of images and symbols—things that connote meaning beyond their literal content, even things by which we are recognizable as Americans. Think of blue jeans, rock 'n' roll, baseball, fast-food, T-shirts, cowboys—or, permitting the symbolism to turn a little darker, think of guns. Foreigners may see us Americans in simple images. We are loud, generous, overbearing, well-intended, violent, naïve, optimistic, rich, ignorant, racist, know-it-alls. Those are only images, of course, but like any image, they may contain some truth.

1. Where did Smokey the Bear come from?

2. A famous patriotic painting shows George Washington crossing a river on Christmas, 1776, foolishly standing up in the boat. The river was the Delaware. Where was Washington going?

3. Which of these best describes the U.S. form of government?

Broad-based kleptocracy _____

Parliamentary democracy _____

Dictatorship of physicians and lawyers _____

Federal democracy _____

Confederacy of sovereign states _____

Confederacy of dunces _____

4. What do Aspen and Telluride, Colorado; Virginia City, Montana, and Virginia City, Nevada; Mammoth Lakes, California; and Tombstone, Arizona, have in common?

5. When the Indian jeans manufacturer Murjani was looking for someone to put her famous name on his tight denim pants, he ended up with socialite Gloria Vanderbilt, thus creating a sappy product called "designer jeans"—an American icon that stormed a world waiting to be told what to buy. But Vanderbilt was his second choice. Who was his first?

6. What do these countries, among others, have in common: Lebanon, the Dominican Republic, Cambodia, Nicaragua, Honduras, Korea, Somalia, Panama, Grenada, and Vietnam?

7. Which state is the Coyote State?

8. *Gone with the Wind* is probably America's most famous movie. Where was it, and the novel by Margaret Mitchell, set?

9. Why is Colorado the Centennial State?

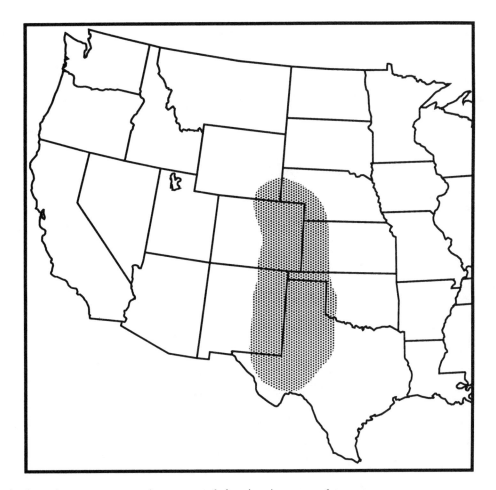

The region marked on this map was at the center of the development of
one of the United States' most persistent and forceful images, among both ourselves and
foreigners. What is it?

Answer: This is where the rodeo began. Originally called "cowboy contests," the
first of these on record was in 1869 in a tiny eastern Colorado town called Deer
Trail. Today, major rodeos can be found as far east as Iowa, as far west as Salinas,
California, and as far north as Calgary, Alberta.

10. While Mark Twain is often associated with Hannibal, Missouri, and life on the Mississippi River, he did much of his writing somewhere else. Where?

11. Sometimes life in America seems like the world's largest merry-go-round, a device which, in fact, is in this country. Where?

12. Disney World, on 28,000 acres the world's largest amusement park, is actually two parks: The Magic Kingdom, and Epcot. Can you say what Epcot means?

13. Where is Boys Town?

14. What distinguishes Minnesota, Iowa, and Oregon as a group from all other states?

15. Can you match the monument with its city?

Seagull Monument	San Francisco
Washington Monument	Lake Wales, Florida
Buckingham Fountain	Liberty Island, New Jersey
U.S.S. *Arizona*	Outside Atlanta
Stone Mountain	Salt Lake City
Bok Tower	Washington, D.C.
Coit Tower	Pearl Harbor, Hawaii
Statue of Liberty	Chicago

16. Where did the American cowboy come from?

17. Where is America's largest circus museum?

18. Where is the Baseball Hall of Fame?

19. Match these American icons with a state listed below:

California	Big Sky
Oregon	Tarheels

North Carolina	Alligators
Florida	Ten Thousand Lakes
Louisiana	Lumberjacks
Minnesota	Cajuns
Montana	Movie stars
Utah	Speedway
Texas	Mormons
Tennessee	Pecos Bill
Indiana	Country music

20. Where might you find the most valuable piece of art in the United States?

21. The bloodiest military battle in the history of the hemisphere occurred in the United States. When and where?

22. Name a hit musical named for a state.

23. What was Benjamin Franklin president of in 1787?

24. What was the hometown of John Steinbeck?

25. Where was the shoot-out at the O.K. Corral?

26. What does this geographical-time sequence—Dallas-1963, Memphis-1968, Los Angeles-1968—signal to you?

27. About 25,000 crimes were committed in 1992 by people dressed up as _____ _____ . Fill in the blanks.

28. You might guess that the average child laughs more than the average adult in America, and you'd be right. But guess how much more by choosing the number of times each laughs daily.

Children 400 and adults 15 _____

Children 150 and adults 65 _____

Children 35 and adults 15 _____

29. Complete this sentence accurately, so that it reveals a shocking truth about America: More _____ men go to _____ than to college.

30. Where is the Johnson Space Flight Center?

31. If you wore a size 6 dress twenty years ago, and you have "filled out" over the years to the point where you probably couldn't get into one of those dresses if you had one, how is it that you can still wear a size 6 off the rack today?

32. Here are two statements by great American political leaders. Can you name the leader who made each?

(a) "Were it left to me to decide whether we should have a government without newspapers, or newspapers without a government, I should not hesitate a moment to prefer the latter."

(b) "Nothing can now be believed which is seen in a newspaper. Truth itself becomes suspicious by being put into that polluted vehicle."

33. Why is it that in Alabama they say, "Thank God for Mississippi," and in Mississippi they say, "Thank God for Alabama"?

34. Disneyland and Disney World have more in common than mouse ears and wholesome images. Can you name something else? (Hint: It may have something to do with climate.)

35. Which of these is true about West Feliciana Parish, just north of Baton Rouge, Louisiana?

It has the highest ratio of men to women of any parish or county in the country. _____
It has the highest ratio of women to men of any parish or county in the country. _____
Explain your answer, if you can.

36. "I have a dream," said Martin Luther King. "I have a dream that one day, on the red hills of Georgia, sons of former slaves and the sons of former slaveowners will be able to sit down together at the table of brotherhood. . . ." Where did King make this eloquent speech?

Answers

Images

1. There was a real Smokey, a cub found badly burned after a forest fire in New Mexico's Lincoln National Forest. The orphan was treated, nursed by a ranger at home, and placed in the National Zoo in Washington, D.C. "Only you can prevent forest fires," read the posters of a national campaign. Smokey, the true bear, died in late 1976. Smokey the Bear was officially retired as the U.S. Forest Service symbol in 1975, though he still appeared on posters from time to time for years thereafter.

2. Trenton, New Jersey.

3. The people's choice is federal democracy.

4. All are mining boomtowns that became ghost towns and which have been redeveloped as tourist towns.

5. Jacqueline Bouvier Kennedy Onassis.

6. The United States has sent military personnel, not always with congressional or popular approval, into all these countries within the last fifty years.

7. South Dakota is the Coyote State.

8. Georgia. Atlanta was Mitchell's hometown.

9. Colorado is the Centennial State because it was admitted to the Union in 1876, when the Union was 100 years old.

10. For thirty years, Twain lived and wrote on Nook Farm in Connecticut.

11. The Columbia, at Great America, an amusement park in Santa Clara, California, is 27.5 feet across,

100 feet high, with two decks of 103 animals to ride on, and spins at a leisurely 7 rpm.

12. Epcot stands for Experimental Prototype Community of Tomorrow.

13. Boys Town is in eastern Nebraska. Founded in 1917 by the famous Father Flanagan, a Roman Catholic priest, it is a home and school for abandoned and neglected boys and, since 1979, girls.

14. These three are the only states without a major military base.

15. The Seagull Monument is in Salt Lake City, the Washington Monument in Washington, D.C., Buckingham Fountain in Chicago, U.S.S. *Arizona* in Pearl Harbor, Stone Mountain outside Atlanta, Bok Tower in Lake Wales, Coit Tower in San Francisco, Statue of Liberty in New Jersey.

16. The technology and habits of the cowboy originated in Mexico, where the word was *vaquero*, source of our term "buckaroo." By the later 1500s, cattle ranching had spread northward from its original heartland in the valley of Mexico. From Texas it spread to Hollywood. From Hollywood, to the world.

17. In Sarasota, Florida.

18. In Cooperstown, New York.

19. California has its movie stars; Oregon, lumberjacks; North Carolina, Tarheels; Florida, alligators; Louisiana, Cajuns; Minnesota, ten thousand lakes; Montana, a big sky; Utah, Mormons; Texas, Pecos Bill; Tennessee, country music; Indiana, the Indi-

anapolis Speedway. Some of these icons are commemorated on license plates.

20. Van Gogh's "Irises" came to reside in the J. Paul Getty Museum in Malibu, California, in late 1993. The picture—of French flowers, not American— was purchased privately, but at its last public auction in 1987 it went for $53.9 million.

21. The Battle of Gettysburg in the early days of July 1863, during the Civil War. The massed forces totaled 150,000 and more than one fourth of them were killed. As it happened, the Northern forces attacked from the South and the Southern forces from the North.

22. *Oklahoma*, by Rodgers and Hammerstein.

23. The country's first abolitionist society, founded in 1775 by Anthony Benezet.

24. Salinas, California, which named the town library after him and preserved the family home. The place which got boutiqued was nearby Monterey, the setting for *Cannery Row* and other Steinbeck novels.

25. Tombstone, Arizona Territory, in 1881.

26. The assassinations of John F. Kennedy, Martin Luther King, and Robert F. Kennedy.

27. Police officers.

28. American children laugh 400 times a day on the average; adults laugh only 15 times.

29. More black men go to jail than to college. More than mere image is going on here, but the image created does great injustice to the many African-American men who work hard and play by the rules.

30. In Houston, as in "Houston, Eagle has landed," which are the first words transmitted from the moon.

31. Maybe the old dress shrank. More likely, you can still wear the size 6 dress even though you're larger because dress manufacturers keep increasing the "sizes" of sizes. "Older women are bigger and fatter, but they still want to be a size 4 or 6 even though they're really an 8 or 10," said Bud Konheim of Nicole Miller Ltd. "We will cut a big size for the older woman and call it a 4 or a 6 and everyone will be happy." Men's suitmakers have not yet learned such thoughtfulness.

32. (a) Thomas Jefferson. (b) Thomas Jefferson.

33. It's a mean, sad joke on both states, which seem to turn up at or near the bottom of every list of states when it comes to quality of life, poor health, and education.

34. Both theme parks are in Orange County—Disneyland in Orange County, California, and Disney World in Orange County, Florida. This cryptic parallelism may be simple coincidence, or it may have something to do with the fact that climates in which citrus fruit thrives are also climates tourists love.

35. West Feliciana Parish has the highest male-to-female ratio in the nation. There are 212 men to every 100 women. But things are not always what they seem, and this richness of maleness would not necessarily make West Feliciana a good place for females to seek a mate. The ratio is the result of its being site of the Angola State Prison, where 5,200 male inmates live. On the other hand, West Feliciana is a very beautiful place, with a thriving tourist trade based on its lovely old plantation houses set amid moss-hung oaks. John James Audubon visited in 1823 and sketched eighty different birds. The lowest ratio of men to women is 78 to 100 in South Boston City, Virginia. (We don't know why.)

36. King's most famous speech was made at the Lincoln Memorial in Washington, D.C., to commemorate the centennial of the Emancipation Proclamation.

Roll Call of the States

Here is a final set of questions, whose answers, simply stated, are states.

1. In what state is the baseball bat known as a Louisville Slugger manufactured?

2. What state gave us Woody Allen, George Burns, Henry James, Groucho Marx, and Herman Melville?

3. Which state was the fifth richest in the nation before the Civil War, but is the poorest today?

4. What state gave us William Henry Harrison, Thomas Jefferson, James Madison, James Monroe, Zachary Taylor, John Tyler, George Washington, and Woodrow Wilson? What do these men have in common?

5. What is the largest state east of the Mississippi?

6. "What the air-conditioner did for the Sunbelt, the jetliner has done for _____ ." Fill in the name of the state.

7. What state gave us these famous "Americans": Ta-sunko-witko, Tatanka Iyotake, and Martha Jane Burk? And by what names do we know them?

8. Which state is the "uranium capital of the world"?

9. Which state's state bird is named after that state's biggest city—or vice versa?

10. What state gave us Bing Crosby and Jimi Hendrix?

11. *Maugh-wau-wa-ma* means "large plains" or "mountains and valleys alternating" in the Delaware Indian language. What state takes its name from this phrase?

12. Which state is the most densely populated?

13. In which state has gambling been the mainstay of the economy since World War II?

14. What state gave us Bill Cosby, Betsy Ross, Andy Warhol, Johnny Weismuller, and Milton S. Hershey?

15. What state is technically, according to its name, a plantation?

16. What state gave us James Agee, Davy Crockett, Aretha Franklin, and Dolly Parton?

17. Which state has the country's largest Scandinavian-American population?

18. Of what state are these among its famous natives: Ezra Taft Benson, Gutzon Borglum, Ezra Pound, Lana Turner?

19. The citizens of which state buy the most books per capita?

20. Which state pioneered many social programs we now take for granted in the rest of the country, including aid to dependent children, workmen's compensation, and old-age assistance?

21. In what state were Studebaker cars built—until 1963?

22. Of which state is the state sport mushing?

23. Which state, once part of another state, was admitted to the Union in its own right in 1863?

24. Which state catches the most lobsters?

25. In what state is the geographic center of the continental United States? (Hint: It's near a town called Lebanon.)

26. Which state gave us Emily Dickinson, Jack Kerouac, Edgar Allen Poe, Henry David Thoreau, Winslow Homer, and George Bush?

27. The name of which state is a corruption of the word *iliniwek,* meaning "land of the superior men," in the language of its natives when the French explorers arrived?

28. What state, called the arsenal of the nation during the American Revolution, was leading the country in defense contract profits per person in 1993?

29. Ten percent of the world's pesticides are used each year in a single U.S. state. Can you name it?

30. In what state was Steve Martin, the actor, writer, and comic, born?

31. In what state do as many as two thirds of the people consider themselves born-again Christians?

32. Which state was the last to ratify the U.S. Constitution—and the last Southern state to secede from the Union?

33. What state is the leading grower of Christmas trees?

34. Which was the first state to permit inheritance taxes to be paid with works of art?

35. What conservative farm state, isolated in the middle of the country, is far from isolationist when it comes to international trade and economics—and why?

36. What state gave us Angie Dickinson, Louis L'Amour, and Peggy Lee?

37. In which state are housing costs highest?

38. What state has the fewest foreign-born residents?

39. The economy of which state is improved by its time zone?

40. Which state has the lowest state taxes?

41. Which state has the highest state taxes?

42. Which state is the richest per capita in personal income?

43. Which state manufactures a product named for an Ottawa Indian chief?

44. Which state has the world's largest concentration of Frank Lloyd Wright's buildings?

45. In which state has the du Pont family enjoyed economic and political prominence for generations?

46. Which state gave us Bob Dylan, F. Scott Fitzgerald, Judy Garland, and Sinclair Lewis, and still produces 60 percent of the nation's iron ore?

47. The most likely place in the world to be murdered is in the United States. It is not exactly a state, though it is a political unit in some ways similar to a state—which is a hint. Where in the United States is it?

48. Which was the only state to create a Martin Luther King holiday by popular vote?

49. Which was the last state to establish a Martin Luther King holiday?

50. Of the top five cities in car theft, four of them are in one state. Name the state.

51. Which three states are the most popular destinations for people in recreational vehicles?

52. In two of these political units live the most physically active people in the United States. Can you pick them out?

    California

    Colorado

    Florida

    Wisconsin

    Maine

    New Mexico

    New Hampshire

    Washington, D.C.

53. Students in ten of these states get the best SAT scores in the country. Pick the best ten:

Iowa	Virginia
Massachusetts	North Dakota
South Dakota	Illinois
Connecticut	Maine
Vermont	New Hampshire
Wisconsin	Idaho
Oregon	California
Maryland	New Jersey
Minnesota	New Mexico
Nebraska	Mississippi

54. In which state is it illegal to stick your tongue out at a fisherman who has caught a smaller fish than you?

55. Which state gave us Muhammad Ali, Colonel Harlan Sanders, and Robert Penn Warren?

56. In which state did the Birdman of Alcatraz pamper his fowl?

57. In which state is one in every six homes heated solely with wood?

58. Which American colony (later a state) was founded by Quakers?

59. Where is the center of Mormon culture in the United States?

60. What state has the greatest proportion of children?

61. What state has the highest proportion of old people?

62. What became of Icebergia, Polaria, and Walrussia?

63. In 1804, Aaron Burr shot and killed Alexander Hamilton in a duel. Where did this happen?

64. In what state is the mythic Field of Dreams—from the movie adaptation of W. P. Kinsella's novel *Shoeless Joe*?

65. Which is America's dairy state?

66. What state leads the country in agricultural production?

67. Where is Gettysburg?

68. What Southern state was the first to secede at the start of the Civil War?

69. In 1876, Lt. Col. George Custer and his Seventh Cavalry were wiped out at the Little Big Horn. Where is that?

70. Where did American women first gain the right to vote?

71. Where was the Ku Klux Klan founded?

72. Where was the shoot-out at the O.K. Corral?

73. Where was the first F. W. Woolworth's Five- and Ten-Cent Store?

74. Where are the offices of the National Geographic Society?

75. Where was the first overhead electric trolley?

76. Where was the first American skyscraper?

77. Where is the hometown of Coca-Cola?

78. What state gave us Louis Armstrong, Truman Capote, Fats Domino, and Lillian Hellman?

79. Where is the home of America's corporate soap giant, Procter & Gamble?

80. Where did the Manhattan Project, named for a well-known island in New York, actually do most of its famous work, and what was that work?

81. In what state was the planet Pluto discovered?

82. Probably the most important American civil rights advance in the twentieth century was the Supreme Court's *Brown* v. *Board of Education*, in 1954, in which separate was declared to be inherently unequal. In which state was the losing board of education?

83. In what state is the Dismal Swamp?

84. In what state is Mount Rushmore?

85. In what state is Mount Vernon?

86. In what state is Mount St. Helens?

87. In what state is Stone Mountain?

88. In what state is the Statue of Liberty?

89. What state has the highest number of abortions per live births?

90. Richard Nixon was elected United States senator in 1950 from what state?

91. In what state is Yoknapatawpha County?

92. Match the states with these National Historic Sites:

Abraham Lincoln Birthplace	California
Carl Sandburg Home	Kentucky
Thomas Edison Birthplace	Georgia
Eugene O'Neill Birthplace	Iowa
Golden Spike	Ohio
Bent's Old Fort	New Jersey
Frederick Douglass Home	North Carolina
Herbert Hoover Birthplace	Colorado
William Howard Taft Birthplace	Utah
Jimmy Carter Birthplace	District of Columbia

93. Which state has the fewest people per square mile?

94. In which state are there "wet liberals" and "dry conservatives"?

95. What state has the most hospital beds per person?

96. Which state has the most physicians?

97. What state has the highest infant mortality rate?

98. What state, in 1990, had the lowest per capita property taxes?

Alabama _____

Mississippi _____

Minnesota _____

Vermont _____

California _____

Florida _____

99. Which state has been called "a vale of humility between two mountains of conceit"?

100. Which state has the most golf courses?

101. Where do people say "bo-it" for boat, "kyar" for car, "gyarden" for garden, and "swimp" for shrimp?

102. What state has the tallest structure in the United States?

103. "Frozen yogurt came from _____ . Tofu did. Jogging came from _____ ." Fill in the state's name to complete the quote.

104. What state is composed entirely of mountain-tops?

105. Which state has no straight-line boundaries?

106. What state pays the highest federal income taxes per capita?

107. What state has the highest percentage of households participating in school lunch and food stamp programs?

108. What state has the highest per capita circulation of daily newspapers?

109. What two states adjoin the Chesapeake Bay?

110. Which two states adjoin no other states?

111. Which state has the most nuclear power plants?

112. Alaska has the highest per capita energy consumption in the nation. That makes sense because Alaska is big and cold, thus requiring much fuel. But which state in the lower forty-eight has the highest per capita energy consumption?

113. In 1989, the most recent year for which statistics are available, which state had the highest death rate?

114. Which state considers itself to be the "Heart of Dixie"?

115. Which state's name means "mainland"—in two languages?

116. Which state is bordered by southern states, midwestern states, and plains states?

117. What state gave us Gary Cooper, Chet Huntley, and Myrna Loy?

118. What state gave us Fred Astaire, Marlon Brando, Loren Eiseley, and Malcolm X?

119. What state gave us Sherwood Anderson, Thomas Edison, Annie Oakley, Art Tatum, and James Thurber?

120. Why would a state bother to have a "state amphibean"?

121. Only one state has a unicameral legislature. Which is it?

122. What state has the most counties?

123. In what state, now presumed to be among the most politically liberal, was the John Birch Society formed?

124. In which state is 90 percent of the country's mohair produced? And what is a mo?

Roll Call of
the States

1. The Louisville Slugger is made in Jeffersonville, Indiana.

2. New York.

3. Mississippi. The Civil War cost Mississippi 30,000 men (65 percent of the Southerners who died in it) and bankrupted the plantation owners when the slaves were freed.

4. Virginia gave us all these presidents.

5. Georgia.

6. Hawaii.

7. South Dakota gave us Crazy Horse, Sitting Bull, and Calamity Jane.

8. New Mexico.

9. Maryland: the Baltimore oriole.

10. Washington gave us these two musicians.

11. Wyoming.

12. New Jersey is the most densely populated state, with fifteen times the national average.

13. Nevada. Gambling produces half the state's tax revenues.

14. Pennsylvania.

15. "Rhode Island and Providence Plantations" is the formal name of this state. It is the smallest state but has the longest name, and its motto is "Hope."

16. Tennessee.

17. Minnesota.

18. Idaho.

19. Montanans buy more books per capita than citizens of any other state.

20. Wisconsin.

21. South Bend, Indiana.

22. Mushing—or dogsledding—is the state sport of Alaska.

23. West Virginia.

24. Maine catches more lobsters than the four other New England states put together. In 1992, Maine lobster revenues totaled $72 million.

25. Kansas.

26. Massachusetts.

27. Illinois.

28. Connecticut.

29. California.

30. Steve Martin was born in Waco, Texas, thus becoming, perhaps, the original wacko from Waco.

(Far from wacko, Martin is a first-rate writer of what might be called serious philosophical comedy, and a serious collector of modern art.)

31. Oklahoma.

32. North Carolina.

33. Oregon is the leading grower of Christmas trees.

34. Maine.

35. Iowa, which exports a fourth of its produce and must stay abreast of world economic affairs.

36. North Dakota.

37. Housing costs are highest in Hawaii, the paradisiacal South Pacific islands where the median price for a house is $335,000. Rents and food costs are also among the highest in the nation. Hawaii ranks twenty-fourth in average income, and many residents, priced out, are moving back to the mainland.

38. Wyoming has the fewest foreign-born residents (only 7,647). Mississippi has the lowest percentage of foreign-borns (0.8).

39. Hawaii's time zone, five hours earlier than New York and two hours earlier than California, allows islanders to do telephone marketing to the mainland during regular business hours and get people on the phone in the evening. Honolulu law firms are often engaged to write legal briefs for New York firms that need them overnight.

40. Alaska, the only state that kicks back a portion of natural resources revenue to residents.

41. New York.

42. Connecticut is the richest state per capita.

43. Michigan makes the Pontiac.

44. Florida has the biggest Wright complex, at Florida Southern College in Lakeland.

45. Delaware.

46. Minnesota. The iron ore comes from the famous Mesabi Range.

47. The most likely place in the world to become a victim of murder or manslaughter is Puerto Rico, according to the World Health Organization. (One is least likely to suffer a violent end in England or Wales, but Northern Ireland is near the top of the list, just below the United States.)

48. Arizona, the very same state whose governor, Evan Mecham, rescinded a King holiday five years before and cost the economy millions in convention revenue and bad publicity.

49. New Hampshire didn't create a King holiday until January 1993.

50. New Jersey.

51. Arizona, Florida, and Texas. Rec-vee owners say they like their way of life for its freedom and economy, and because it has all the comforts of home.

52. The most active Americans live in Colorado and New Hampshire. The most sedentary live in our nation's capital, which may not come as a surprise, since that's where Congress is. Fifty-two percent of Washingtonians said they had engaged in zero leisure-time physical activity in the past month. (That slander against Congress probably isn't fair; many Congress members live outside the district.)

53. The states in the first column are the top ten.

54. Alaska; so there.

55. Kentucky.

56. Robert Franklin Stroud kept up to twenty birds in his isolation cell at Leavenworth Penitentiary, in Kansas, until 1942. Then, to suppress and punish the fame he had earned with the media, the prison system transferred him to Alcatraz, in San Francisco Bay, and forbade him to keep pets.

57. Vermont.

58. Pennsylvania.

59. Salt Lake City, Utah, is the center of Mormon life, but the revelation of Joseph Smith occurred near Palmyra, New York, on top of a hill built by glaciers during the last Ice Age.

60. Utah has the greatest proportion of children, because Mormons tend to believe in large families. The Alaskan population is also very young.

61. Florida has the highest proportion of elderly people. In many Florida counties, more than 20 percent of the people are over sixty-five.

62. These were early names given to Alaska by the many who jeered at the Alaska purchase from Russia in 1867. The jeers evaporated as the purchase revealed itself to have been a good deal.

63. Hamilton had his choice of weapons and site, and he chose (to die in, as it turned out) Weehawken, New Jersey.

64. Iowa.

65. Wisconsin—in most years it says so on the license plates. However, California now produces more milk.

66. California, by bunches, bushels, and bucks.

67. The site of the most famous, deadliest, and probably most important, battle of the Civil War is in south-central Pennsylvania. Gettysburg, subject of a fine novel, *The Killer Angels,* by Michael Schaara, remains the bloodiest battle ever fought in the United States.

68. South Carolina, site of the first battle (at Fort Sumter) seceded first. When the war began in 1861, the North had nearly four times as many people as the South, which makes it surprising that hostilities lasted as long as they did. In the end, disease killed twice as many as hostile fire.

69. The battle site, along the Little Big Horn River, is in southeastern Montana.

70. American women first voted in the Wyoming Territory, in 1869.

71. In Pulaski, Tennessee, in 1866.

72. The O.K. Corral was in Tombstone, in the Arizona Territory. The shoot-out occurred in 1881.

73. The first Woolworth's opened in Lancaster, Pennsylvania, in 1879.

74. The society's offices are in Washington, D.C.

75. In Richmond, Virginia, it began running in 1888.

76. The first iron and steel frame skyscraper was the Home Life Insurance Building in Chicago, ten stories high and built in 1885. Skyscrapers were made possible by the invention, by Otis, of the elevator.

77. Coca-Cola, which first went on sale (containing cocaine) in 1866, is very high on Atlanta, Georgia, which the company calls home.

78. Louisiana.

79. Procter & Gamble is in Cincinnati, Ohio. James Gamble and Harley Procter got their start in 1879 with Ivory Soap.

80. The Manhattan Project developed the world's first atomic bomb in Los Alamos, New Mexico.

81. Pluto was discovered by astronomer Clyde Tombaugh working at Arizona's Lowell Observatory in 1930.

82. In Topeka, Kansas, which previously had a legally segregated school system.

83. The Dismal Swamp, which is not all that dismal though it is indeed a swamp, is in eastern Virginia south of Norfolk.

84. Mount Rushmore, a wonderful hunk of granite kitsch featuring the faces of Lincoln, Washington, Jefferson, and Teddy Roosevelt, is in the Black Hills of South Dakota.

85. Mount Vernon, the estate of George Washington, is on the Potomac River in Virginia.

86. Mount St. Helens is in Washington.

87. Stone Mountain, featuring the mounted figures of Confederate heroes carved in granite and even kitschier than Mount Rushmore, is just outside Atlanta in Georgia.

88. The Statue of Liberty, on Liberty Island in New York Harbor, is in New Jersey.

89. New York.

90. California sent Nixon to the Senate. He was elected president in 1968 and resigned in disgrace in 1974.

91. William Faulkner's fictional county lives in Mississippi.

92. Lincoln was born in Kentucky, Sandburg in North Carolina, Edison in New Jersey, O'Neill in California, the Golden Spike is in Utah, Bent's Old Fort is in Colorado, Douglass was born in the District of Columbia, Hoover in Iowa, Taft in Ohio, Carter in Georgia.

93. Alaska has the fewest, with not even one per square mile. New Jersey, the most densely populated state, has 1,142 persons per square mile.

94. Washington, where the liberals dominate in urban areas west of the Cascades, where it rains all the time, and conservatives dominate the rural, semi-arid areas east of the Cascades, where not long ago, during the ancient-forest-and-spotted-owl controversy, bumper stickers read: SAVE A LOGGER; SHOOT AN OWL.

95. North Dakota has the most beds; Nebraska is second; South Dakota third; showing a regional pattern.

96. Maryland has the most doctors. And the District of Columbia, right next door, has twice as many.

97. Georgia's is highest, at 12.6 per 1,000 live births. Vermont is lowest with 6.8. The District of Columbia, not a state, has an astonishingly high rate of 23.2, which compares to several Third World countries.

98. Alabama had the lowest, averaging $163 per person. Alaska's, at $1,246, was highest.

99. North Carolina, between Virginia and South Carolina.

100. Florida has the most, with 1,100. Golf courses are intense-use creations. No growing surface on earth—not a cornfield nor a suburban lawn nor a botanical garden—requires as much water, pesticide, herbicide, and fertilizer as a golf course's putting green.

101. In and near Charleston, South Carolina.

102. A TV tower in Blanchard, North Dakota, is the nation's tallest structure. It is 2,063 feet high.

103. California.

104. Hawaii. All the islands are the tips of great volcanic peaks rising from the Pacific Ocean floor.

105. Hawaii.

106. Connecticut pays the most, by a wide margin.

107. Mississippi.

108. New York, although the state with the greatest number of newspapers is Texas.

109. Virginia and Maryland are on the Chesapeake.

110. Only Alaska and Hawaii stand alone. Maine borders only one other state.

111. Illinois has the most nuke plants, with thirteen in action in 1989. Pennsylvania had eight.

112. Louisiana. But Louisiana is warmish and smallish, you may be saying—so how does that figure? Well, Louisiana has the nation's largest concentration of petrochemical plants, which devour fuel voraciously.

113. Missouri. Why? We don't know why.

114. Alabama is called the Heart of Dixie.

115. Alaska's name means "mainland" in both Aleut (*alaska*) and in Eskimo (*alakshak*).

116. Missouri.

117. Montana.

118. Nebraska.

119. Ohio.

120. Ask New Hampshire, where the designated creature is the spotted newt.

121. Nebraska has a single-house, or unicameral, legislature. Several other states are considering the idea.

122. Texas, with 254. Altogether, the United States has 3,070 counties. Thirty-one states have counties named for George Washington; 115 counties, such as Prince George, have names memorializing British nobility.

123. The Birch Society, named by founder Robert Welch for a U.S. Army officer killed by Chinese Communists a week after the end of World War II, was founded in Belmont, Massachusetts, in 1957.

124. Ninety percent of the nation's mohair—the soft, silky wool of angora goats—is produced in Texas. Texas also produces 20 percent of the nation's wool. There is no mo.